Nikola Tesla

The Great Truth About Gravity

(The Life of a Genius and the Impact of His Work & a Scientific Biography)

Jimmy Howard

Published By **Tyson Maxwell**

Jimmy Howard

Nikola Tesla: The Great Truth About Gravity (The Life of a Genius and the Impact of His Work & a Scientific Biography)

ISBN 978-1-7770663-5-2

Legal & Disclaimer

The information contained in this book is not designed to replace or take the place of any form of medicine or professional medical advice. The information in this book has been provided for educational & entertainment purposes only.

The information contained in this book has been compiled from sources deemed reliable, and it is accurate to the best of the Author's knowledge; however, the Author cannot guarantee its accuracy and validity and cannot be held liable for any errors or omissions. Changes are periodically made to this book. You must consult your doctor or get professional medical advice before using any of the suggested remedies, techniques, or information in this book.

Upon using the information contained in this book, you agree to hold harmless the Author from and against any damages, costs, and expenses, including any legal fees potentially resulting from the application of any of the information provided by this guide. This disclaimer applies to any damages or injury caused by the use and application, whether directly or indirectly, of any advice or information presented, whether for breach of contract, tort, negligence, personal injury, criminal intent, or under any other cause of action.

You agree to accept all risks of using the information presented inside this book. You need to consult a professional medical practitioner in order to ensure you are both able and healthy enough to participate in this program.

Table Of Contents

Chapter 1: A Genius

You are probably the most extensively recognized inventor of the twenty first century, but inside the twentieth now not hundreds, you had been regularly ridiculed or even smeared for being a recluse and, given the reports of your love of that "white" pigeon, a piece of a crack pot as properly. What do you're making of such bad thoughts concerning your statistics and effect on humanity?

The cutting-edge development of man is vitally depending on invention. It is the most vital crafted from his modern thoughts. Its very last reason is the complete mastery of thoughts over the fabric international, the harnessing of the forces of nature to human desires. This is the tough mission of the inventor who's frequently misunderstood and unrewarded. But he finds enough compensation inside the appealing wearing activities of his powers and inside the facts of being one in every of that pretty privileged elegance with out whom the race may have extended inside the past perished within the bitter war in opposition to pitiless elements.

Speaking for myself, I definitely have already had extra than my complete diploma of this first rate entertainment, so much that for many years my life changed into little quick of non-forestall rapture. I am credited with being one of the hardest human beings and in all likelihood I am, if belief is the identical of tough paintings, for I even have committed to it nearly all of my waking hours. But if paintings is interpreted to be a genuine primary performance in a real time steady with a inflexible rule, then I can be the worst of idlers. Every try beneath compulsion needs a sacrifice of life-energy. I never paid this form of fee. On the opposite, I even have thrived on my thoughts.

And what do you are saying to the Pigeon rumor?

Well, I can high-quality say that it's been claimed that I come to be in love with a white pigeon as a good buy as a person cherished a woman, however that is heresy. This is a rumor to discredit me as an lousy lot as each other horrific remarks that have portrayed me to the hundreds as one which has grow to be dissociated with society. I can't communicate to who or why every person may say the type of thing.

What I did declare turn out to be that the white pigeon visited me via an open window at my inn, and I take delivery of as actual with the hen had come to tell me a few element profound. When it sat on the sill, I noticed powerful beams of moderate inside the fowl's eyes, it emerge as a real slight, a powerful, brilliant, blinding mild, a light extra excessive than I had ever produced with the useful resource of the only lamps in my laboratory. The pigeon then died in my hands, and I knew in that second, I knew, I had completed my life's paintings.

Do you positioned there will ever be a manner to the due to this of existence?

Of all of the endless fashion of phenomena which nature offers to our senses, there's none that fills our minds with extra wonder than that inconceivably complicated movement, inner its entirety, we designate as human existence; Its mysterious starting place is veiled in the all of the time impenetrable mist of the past, its man or woman is rendered incomprehensible via its infinite complex, and its vacation spot is hidden within the unfathomable depths of the destiny. From in which does it come? What is it? Whither does it generally will be inclined? These are the first-rate questions which all the sages of all stated time have endeavored to answer and nary a one has executed.

Ok, so that you're pronouncing there can be no manner to understand the actual which means to Life or even looking for the foundation of Life has examined fruitless.

What does your technological expertise tell us about information life?

That is actual to a degree. We may additionally moreover moreover in no way apprehend the that means or locate the beginning of life but with a revolutionary attitude we are capable of always refine our know-how of Life. This is the growth and modern improvement of all life. To look at, to examine, to adapt and via doing so, we increase without limits because thinking about the fact that Life is in reality countless, there's no give up to the idea of increase. Growth is endless.

In the feel that I speak of Life and my perception of our global is from my scientific angle, that is the manner I modified into raised, along thing the technology in which I changed into raised in concept, schooling in terms of societal improvements. And this is furthermore how I generalize the understanding round me. With that stated, current-day-day technological know-how says the solar is the beyond, the earth is the prevailing and the moon is the future. From an incandescent mass we've got originated and right proper right into a frozen mass we're in a position to show. Merciless is the regulation of nature and swiftly and irresistibly we are inquisitive about our doom. It has been stated that the sun will shine brilliant no more and in its place handiest a dark isolated remnant shall remain.

However, there are others who get hold of as actual with the spark of life can in no way be outstanding, but will constantly tour and attempting to find the proper habitat for growth and improvement. Professor Dewar's lovely experiments with liquid air, which show that germs of natural existence are not destroyed by way of bloodless, regardless of how excessive; consequently they will be transmitted via the interstellar cosmos. Meanwhile the cheering lighting of technological understanding and art and precise extracurricular enjoyment, ever growing in depth, cast off darkness from our course, and marvels they reveal, and the enjoyments they provide, make us measurably forgetful of the opportunity of a gloomy future.

So, steady with you, we will never understand the muse or that means of Life in its entirety, however our facts of Life becomes extra easy with a normal try at commentary and gaining knowledge of?

Correct. Though we may also in no manner be able to understand human lifestyles, we apprehend with absolute fact, that lifestyles is countless motion. Constant, everlasting motion however the nature of this is unknowable. But the existence of motion always implies a frame it really is being moved and a pressure that is moving it. Hence, everywhere there may be existence, there can be a mass moved with the resource of a pressure. All mass possesses inertia, all pressure has a tendency to persist. Owing to this ordinary assets and situation, a frame, be it at relaxation or in movement, has a bent to stay in the same america of the usa, and a pressure, manifesting itself everywhere and via a few thing purpose, produces an identical opposing pressure, and as an absolute necessity of this it follows that each movement in nature must be rhythmical.

Long inside the beyond this simple reality become in truth noted through Hebert Spencer, who arrived at it via a extremely unique approach of reasoning. It is borne out inside the good buy we recognize—within the motion of a planet, inside the surging and ebbing of the tide, within the reverberations of the air, the swinging of a pendulum, the oscillations of an electric powered powered powered cutting-edge-day, and inside the infinitely severa phenomena of natural existence. Does now not the complete of human lifestyles attest to it? Birth, increase, antique age, and shortage of lifestyles of an individual, family, race, or u.S.A. Of the united states, what is all of it but a rhythm? All life-manifestation, then, even in its maximum hard form, as exemplified in guy, but involved and inscrutable, is exceptional a movement, to which the same desired laws of motion which govern at a few stage within the bodily universe have to be applicable.

But how has the person of guy interfered with or otherwise affected the growth and development of lifestyles inside the world?

When we communicate of guy, we have got a concept of humanity as a whole, and earlier than making use of medical strategies to, the investigation of his motion we must take transport of this as a bodily reality. But can genuinely everyone doubt these days that each one the billions of people manufactured from all the innumerable kinds and characters constitute an entity, a unit? Though unfastened to suppose and act, we're held together, much like the celebrities in the firmament, with ties inseparable. These ties can't be seen, however we can revel in them. I reduce myself within the finger, and it pains me: this finger is a part of me. I see a pal damage, and it hurts me, too: my friend and I are one. And now I see down an enemy, a lump of matter quantity which, of all of the lumps of matter within the universe, I care least for, and it nevertheless grieves me. Does this no longer display that every parents is satisfactory part of an entire?

For a long time this concept has been proclaimed within the consummately smart teachings of religion, possibly now not by myself as a way of insuring peace and concord amongst guys, however as a deeply based fact. The Buddhist expresses it in a unmarried manner, the Christian in a few exclusive, but each say the same: We are all one.

Metaphysical proofs are, but, no longer the best ones which we are capable of bring about in guide of this idea. Science, too, recognizes this connectedness of separate individuals, even though not pretty inside the identical feel as it admits that the suns, planets, and moons of a constellation are one frame, and there can be no doubt that it's going to in all likelihood be experimentally showed in times to return, even as our technique and techniques for investigating psychical and unique states and phenomena shall had been brought to awesome perfection.

Still greater: this one person lives on and on. The individual is ephemeral, races and global places come and bypass away, however man remains. Therein lies the profound distinction a few of the individual and the entire. Therein, too, is to be located the partial clarification of loads of those sudden phenomena of heredity which might be the prevent result of countless centuries of feeble however continual have an effect on.

Furthermore, if guy became to also flow extinct, likely with the aid of the usage of the collective folly of a race, the earth shall over again pass again lower again to its untouched beauty and the natural laws of increase and improvement. It is honestly internal man's electricity to move again to this state right now, the identical country that our ancestors as soon as roamed, but with the invention of fireplace, there in that second became hooked up a temperment of growth of a modern day type; therein sparked the economic growth, which in no manner ceases till paradoxically, the very last mild dims. And it will dim on guy, if he persists in tough the may additionally of Life. If man keeps to fight towards the herbal increase of his domestic, and fails to understand the damaging route wherein commercial agency development requires, he is going to stay no greater however Life shall pass back to all its superb beauty and the climate shall over again skip returned to its herbal state of giving Life and sustenance even as moreover receiving, with the resource of divine order, lifestyles and sustenance in pass back.

So how must you advise guy returns again to our roots and our area inside the herbal order of existence? How are we able to opposite the horrible and poor path you simply defined?

First we want to conceive guy as a mass recommended on via way of a pressure, just like planets and stars inside the direction of the universe. Though this motion isn't of a translatory character, implying a exchange of region, but the popular felony guidelines of mechanical motion are applicable to it, and the power related to this mass can be measured, according with famous principles, via 1/2 of the made of the mass with the square of a positive tempo. So, as an instance, a cannon-ball this is at rest possesses a certain amount of power within the shape of heat, which we degree in a comparable way. We recollect the ball to consist of innumerable minute debris, referred to as atoms or molecules, which vibrate or whirl spherical each different. We determine their hundreds and velocities, and from them the energy of each of these minute systems, and consisting of all of them together, we get an idea of the entire heat-strength contained within the ball, that is best seemingly at rest. In this in easy terms theoretical estimate this strength may additionally moreover then be calculated with the useful resource of multiplying half of of the whole mass—that is half of of the sum of all of the small loads—with the square of a tempo this is decided from the velocities of the separate particles. In like manner we may moreover conceive of human power being measured with the resource of half of the human mass accelerated with the rectangular of the speed which we aren't yet capable of compute. But our deficiency in this expertise will no longer vitiate the truth of the deductions I shall draw, which rest on the business enterprise foundation that the equal laws of mass and stress govern in the direction of nature.

Man, however, isn't an ordinary mass, which includes spinning atoms and molecules, and containing clearly heat-energy. He is a mass possessed of fantastic better capabilities by way of the usage of manner of cause of the progressive principle of lifestyles with which he is endowed. His mass, due to the reality the water in an ocean wave, is being continuously exchanged, new taking the area of the antique. Not pleasant this, however he grows, propagates, and dies, for this reason changing his mass independently, every in bulk and density.

What is maximum exceptional of all, he is able to growing or diminishing his tempo of movement through the mysterious power he possesses thru the usage of appropriating greater or masses much less power from outstanding substances, and turning it into reason strength. But in any given 2d we may furthermore ignore those slow adjustments and count on that human power is measured by half of of of the manufactured from guy's mass with the square of a certain hypothetical speed. However we can also moreover compute this tempo, and some thing we can also take because the same antique of its diploma, we need to, in harmony with this concept, come to the perception that the notable trouble of technological know-how is, and normally is probably, to increase the strength consequently defined.

Chapter 2: The Three Ways Of Increasing Human Energy

Let, then, in diagram a, M represent the mass of guy. This mass is impelled in a unmarried direction thru a pressure f, this is resisted via each other partially frictional and partly terrible strain R, acting in a course precisely opposite, and retarding the motion of the mass. Such an detrimental force is found in each movement and want to be taken into consideration. The difference between the ones forces is the effective stress which imparts a pace V to the mass M in the path of the arrow on the line representing the pressure f. In accordance with the previous, the human strength will then get maintain of via manner of the product ½ MV2 = ½ MV x V, wherein M is the complete mass of guy within the normal interpretation of the term "mass," and V is a fine hypothetical pace, which, inside the present usa of technology, we are no longer able to outline and determine.

To increase the human power is, consequently, identical to growing this product, and there are, as will without issues be seen, first-rate three strategies possible to accumulate this end end result, which can be illustrated within the above diagram. The first manner established in the pinnacle discern, is to increase the mass (as indicated via the dotted circle), leaving the 2 opposing forces the equal. The second manner is to lessen the retarding stress R to a smaller cost r, leaving the mass and the impelling force the equal, as diagrammatically proven in the middle discern. The 1/3 manner, this is illustrated within the closing determine, is to boom the impelling strain to a better price F, at the same time as the mass and the retarding stress R stay unaltered. Evidently regular limits exist as regards boom of mass and cut price of retarding pressure, however the impelling pressure can be increased indefinitely. Each of these 3 possible answers offers a unique trouble of the number one problem of growing human electricity, that is as a consequence divided into three first rate issues, to be successively considered.

That emerge as very well said, albeit very tough to comply with at times. I understand what you're pronouncing for the maximum element. There are essentially handiest three strategies of making modern alternate; (1.) Increase mass to be more than the opposing pressure and your wonderful movement thereby setting up regular fine increase. (2) lessen the opposing strain retaining you once more so to talk, while maintaining steady mass and first-rate motion. (three) developing the affect of your exceptional motion while keeping everyday mass and pulling a ways from the opposing pressure.

With that said, do you receive as actual with guy will find out the way to maintain in mind and improve the long time consequences of such negative behavior and skip returned to the strategies of our ancestors; through manner of using your advanced technology at the equal time as all over again leaving a limited carbon footprint?

Viewed normally, there are very definitely most effective two methods of in reality growing the efficient mass of mankind: first, with the resource of helping and preserving the ones forces and conditions which will be inclined to growth it; and, 2d, via the use of opposing and lowering the ones which have a propensity to diminish it. From an individual mind-set, the mass of the individual will unequivocally be stepped forward thru careful hobby to health, with the aid of sizable food, with the beneficial useful resource of moderation, with the resource of regularity of behavior, via merchandising of familial commitment, via conscientious hobby to children, and, usually stated, by the use of the observance of all the many precepts and legal guidelines of hygiene. Even a proper away awareness on one's expertise of their non secular ideals might also have a number one impact on man or woman increase and development.

But in consisting of new mass to the vintage, 3 instances over again gift themselves. Either the mass introduced is of the equal velocity because the antique, or it is of a smaller or of a higher tempo. To gain an idea of the relative importance of those instances, remember a train composed of, say, 100 locomotives on foot on a song, and assume that, to boom the energy of the moving mass, four extra locomotives are delivered to the educate. If the ones four flow into on the identical pace at which the train goes, the overall energy might be improved 4 in line with cent.; if they will be transferring at handiest one 1/2 of of that tempo, the increase will amount to first-rate one in step with cent.; if they're shifting at times that pace, the increase of energy might be 16 in line with cent. This easy example shows that it's far of best significance to feature mass of a higher speed.

So that specialize in growing the person mass and tremendous motion will in the long run improve the overall mass of mankind?

That’s proper. If we hobby on personal development, in my view first, and then pay unique hobby to the right improvement of our kids, we are capable of see drastic modifications from this period to the subsequent. Societal improvement will develop extra than proportionately, it'll develop exponentially. In reality, it's going to develop in methods that is probably implausible to even fathom.

Stated extra to the factor, if, as an example, the kids can be of the identical diploma of development and enlightenment due to the fact the parents,that is, mass of the "same velocity,"the strength will really boom proportionately to the quantity brought. If they're an awful lot much less clever or advanced, or mass of "smaller pace," there may be a totally slight advantage within the strength; but if they'll be further superior, or mass of "higher tempo," then the brand new era will add very extensively to the sum fashionable of human strength. Any addition of mass of "smaller tempo," beyond that important quantity required by way of using the law expressed in the proverb, "Mens sana in corpore sano,(healthful thoughts in a healthful frame) " ought to be strenuously opposed. For instance, the mere improvement of muscle, as aimed toward in a number of our schools, I keep in mind same to inclusive of mass of "smaller tempo," and I could not commend it, despite the fact that my views were particular once I end up a scholar myself. Moderate workout, making sure the right balance between mind and body, and the very best ordinary performance of overall performance, is, of path, a high requirement. The above instance shows that the maximum important surrender quit end result to be attained is the education, or the increase of the "speed," of the mass newly introduced.

What do you think is the reason of the resistance or the retarding pressure strolling in the direction of humanity and in the end, the man or woman?

Well, it scarcely want be said that the whole lot this is in competition to the classes of faith, spirituality and the laws of hygiene is tending to decrease the mass. For instance, whisky, wine, tea, espresso, tobacco, and other such stimulants are chargeable for the shortening of the lives of many, and ought to be used with moderation. But I do no longer expect that rigorous measures of suppression of behavior positioned through many generations are commendable. It is wiser to educate and pontificate moderation in preference to abstinence. We have emerge as acquainted with the ones stimulants, and if such reforms are to be effected, they should be sluggish and sluggish. Those who are devoting their energies to such ends can also want to make themselves an extended manner greater useful thru turning their efforts in extraordinary pointers, as, for example, closer to supplying herbal water.

Pure water? What do you advocate?

For everyone who perishes from the outcomes of a stimulant, at least 1000 die from the consequences of consuming impure water. This valuable fluid, which every day infuses new life into us, is likewise the chief vehicle via which sickness and loss of lifestyles input our our our bodies. The germs of destruction it conveys are enemies all of the more horrible as they perform their deadly work unperceived. They seal our doom whilst we live and revel in.

Even in this modern age there are plenty and masses of human beings in undeveloped international locations, lots of which can be younger and aged, lack of lifestyles from impure consuming water. What a tragedy that we stay with such upgrades and no matter the fact that a person someplace is death in this very minute from lack of water!

The majority of humans are so ignorant or careless in consuming water, and the effects of this are so disastrous, that a philanthropist can scarcely use his efforts better than thru endeavoring to enlighten individuals who are therefore injuring themselves. By systematic purification and sterilization of the ingesting water the human mass can be very substantially multiplied. It ought to be made a inflexible rule which is probably enforced via law to boil or sterilize the consuming water in each own family and public region. The mere filtering does not manage to pay for enough safety towards contamination. All ice for internal makes use of need to be artificially organized from water very well sterilized. The significance of having rid of germs of ailment from the metropolis water is usually diagnosed, however little is being finished to improve the prevailing situations, as no first-rate technique of sterilizing wonderful portions of water has but been brought ahead. By advanced electrical home equipment we're now enabled to supply ozone cost efficaciously and in big portions, and this perfect disinfectant seems to offer a glad choice to the important question.

Ok, I apprehend. And in phrases of societal vices? What are your thoughts on correcting this wayward trajectory inner society?

Gambling, industrial company rush, and delight, in particular on the exchanges, are reasons of a bargain mass reduce rate, all of the more so due to the truth the people concerned represent gadgets of higher price however it's miles a continuously fluctuating higher rate and in time also can become a terrible variable. Hence, inconsistency with any movement of ordinary durations will constantly thwart the perceived beforehand improvement. Consistency consequently, is vital to private and societal improvement. This essentially results in one's disability of looking at the primary signs and symptoms and signs of an contamination, and careless neglect of the equal, are essential factors of mortality. In noting cautiously each new signal of coming near danger and chance, and making fastidiously each feasible try to stop it, we aren't first-class following wise prison tips of hygiene in the hobby of our properly-being and the fulfillment of our labors, however we are also complying with a better moral responsibility.

Everyone should don't forget his body as a valuable present from one whom he loves in general, as a outstanding paintings of artwork, of indescribable beauty and mastery beyond human concept, and so sensitive and frail that a phrase, a breath, a look, nay, a idea, may additionally moreover injure it. Uncleanliness, which breeds sickness and death, is not most effective a self damaging but a mainly retarding dependancy. In doing our detail to preserve our our bodies loose from infection, healthful, and natural, we're expressing our reverence for the immoderate principle with which they are endowed. He who follows the precepts of hygiene in this spirit is proving himself, up to now, certainly spiritual. Laxity of morals is a terrible evil, which poisons every mind and body, and that's liable for a awesome good buy of the human mass in some global places.

However, a number of the winning customs and dispositions are green of comparable hurtful outcomes. For example, the society existence, present day training and the pastimes of ladies, tending to attract them away from their circle of relatives obligations and make men out of them, need to detract from the elevating awesome they represent, decrease the ingenious modern strength, and motive sterility and a present day weakening of the race. A thousand great evils is probably said, but all prepare, of their bearing upon the trouble below dialogue, they couldn't equal a single one, the want of food, brought on by way of the use of poverty, destitution, and famine. Millions of people die every one year for want of meals, therefore preserving down the mass. Even in our enlightened businesses, and but the numerous charitable efforts, that is nevertheless, in all opportunity, the chief evil. I do now not propose right right here absolute want of meals, but want of wholesome nutriment.

So in vicinity of focusing at the fluctuating variables of the higher and decrease values, which in itself is reason for irregularities and welcoming possibly horrible results, we need to as an alternative awareness our efforts on the powerful nutrients one consumes as a basis for non-public boom?

Correct, a way to offer accurate and extensive food is the maximum crucial query of the day. On a macro scale, the general concepts of elevating livestock as a way of imparting meals is objectionable, due to the fact, inside the enjoy interpreted above, it need to certainly normally normally have a tendency to the addition of mass of a "smaller pace." It is absolutely most first rate to raise veggies, and I think, consequently, that vegetarianism is a commendable departure from the installed barbarous addiction. That we will subsist on plant food and perform our art work even to advantage isn't always a idea, however a nicely-showed truth. Many races dwelling nearly exclusively on greens are of superior body and energy. There isn't any doubt that a few plant food, together with oatmeal, is extra less expensive than meat, and superior to it in regard to both mechanical and highbrow performance. Such food, moreover, taxes our digestive organs decidedly loads much much less, and, in making us extra contented and sociable, produces an quantity of suitable difficult to estimate. In view of those records every attempt ought to be made to forestall the wanton and cruel slaughter of animals, which have to be poor to our morals. To free ourselves from herbal tendencies and appetites, which maintain us down, we need to begin at the very root from which we spring: we have to have an effect on a thorough reform within the individual of the meals.

There seems to be no philosophical necessity for food. We can conceive of prepared beings dwelling with out nourishment, and deriving all the strength they want for the overall performance of their lifestyles capabilities from the ambient medium. In a crystal we've got got got the clean evidence of the lifestyles of a formative life-principle, and although we can not apprehend the lifestyles of a crystal, it's far but a living being. There can be, except crystals, special such individualized, fabric systems of beings, in all likelihood of gaseous constitution, or composed of substance nonetheless greater tenuous. In view of this possibility, nay, possibility, we cannot apodictically deny the existence of prepared beings on a planet truly because of the reality the conditions on the same are wrong for the lifestyles of lifestyles as we conceive it. We can not even, with immoderate super guarantee, assert that a number of them may not be gift right right here, in this worldwide, inside the very midst oldsters, for his or her constitution and existence-manifestation may be such that we aren't capable of apprehend them. In fact, the magical hobby that many swear to have witnessed, may very well assist this point.

Are you suggesting society must vicinity more hobby on developing farming and vegetable boom in an effort to surpass non-meat intake in preference to growing productivity of our farm animals farming?

Yes, the manufacturing of "synthetic food" as a way for causing an growth of the human mass manifestly indicates itself, but a right away attempt of this type to offer nourishment does not seem to me rational, at least no longer for the winning. Whether we must thrive on such food may be very doubtful. We are the give up end result of a long time of non-prevent model, and we can not considerably trade with out unexpected and, in all opportunity, disastrous results. So uncertain an take a look at should now not be attempted. By an prolonged way the brilliant way, it seems to me, to fulfill the ravages of the evil, might be to discover methods of developing the productiveness of the soil. With this item the safety of forests is of an significance which can not be puffed up, and on this connection, moreover, the use of water-electricity for features of electrical transmission, dishing out in lots of methods with the want of burning wood, and tending thereby to wooded location renovation, is to be strongly encouraged. But there are limits inside the improvement to be affected on this and comparable strategies.

To increase materially the productiveness of the soil, it must be more effectively fertilized with the aid of synthetic approach. The query of food-production resolves itself, then, into the query how first-rate to fertilize the soil. What it is that made the soil remains a mystery. To deliver an reason for its basis might be same to explaining the starting vicinity of lifestyles itself. The rocks, disintegrated with the beneficial resource of moisture and warmth and wind and climate, were in themselves no longer capable of keeping lifestyles. Some unexplained scenario arose, and a few new principle got here into effect, and the primary layer able to sustaining low organisms, like moses become shaped. These, by means of the use of their life and loss of life, brought greater of the existence keeping high-quality to the soil, and better organisms can also want to then subsist, and so on and on, till at very last quite advanced plant and animal lifestyles can also want to flourish. But although the theories are, even now, not in settlement as to how fertilization is effected, it's miles a reality, best too properly ascertained, that the soil can not indefinitely preserve life, and a few way have to be discovered to deliver it with the materials which have been abstracted from it by means of way of the vegetation. The chief and most treasured the various ones materials are compounds of nitrogen, and the cheap production of these is, consequently, the vital trouble for the solution of the all-essential food hassle. Our environment includes an inexhaustible amount of nitrogen, and can we but oxidize it and bring those compounds, an incalculable gain for mankind may examine.

But how are we able to increase the charge of the soil if we don't understand the approach of revitalizing soil?

Long inside the beyond this concept took a effective preserve on the creativeness of scientific guys, but an inexperienced way for carrying out this stop end result could not be devised. The hassle became rendered enormously tough by using the outstanding inertness of the nitrogen, which refuses to mix irrespective of oxygen. But right here strength involves our useful resource: the dormant affinities of the detail are awoke with the useful aid of an electric powered powered present day-day of the proper first rate. A lump of coal which has been in contact with oxygen for loads of years without burning will combine with it on the identical time as as speedy as ignited, so nitrogen, excited with the resource of electricity, will burn.

I did no longer be successful, however, in generating electric discharges interesting very correctly the atmospheric nitrogen till a relatively present day date, even though I showed, in May, 1891, in a scientific lecture, a very precise shape of discharge or electric powered flame named "St. Elmo's hotfire," which, besides being capable of producing ozone in abundance, additionally possessed, as I stated on that occasion, fairly the great of interesting chemical affinities. This discharge or flame have come to be then handiest 3 or four inches lengthy, its chemical motion turned into likewise very feeble, and consequently the method of oxidation of nitrogen changed into wasteful. How to intensify this action come to be the question. Evidently electric powered currents of a unusual type had to be produced which will render the technique of nitrogen combustion extra green.

The first improve became made in ascertaining that the chemical interest of the release modified into very drastically extended via the usage of currents of extremely immoderate frequency or fee of vibration. This was an important development, but sensible issues quickly set a particular limit to the improvement on this course. Next, the effects of the electric strain of the modern-day-day impulses, of their wave-shape and special characteristic functions, had been investigated. Then the have an effect on of the atmospheric strain and temperature and of the presence of water and one-of-a-kind our our bodies have become studied, and as a quit result the top notch conditions for causing the most immoderate chemical movement of the discharge and securing the exceptional performance of the method have been steadily ascertained. Naturally, the enhancements have been now not short in coming; although, grade by grade, I advanced. The flame grew big and massive, and its oxidizing motion grew extra severe. From a mere brush-discharge a few inches prolonged it advanced into a incredible electric phenomenon, a roaring blaze, devouring the nitrogen of the surroundings and measuring sixty or seventy toes across. Thus slowly, nearly imperceptibly, opportunity have turn out to be accomplishment. All isn't always however performed, via any method, however to what a degree my efforts were provided an concept may be acquired from an inspection of Fig. 1 (p. 176), which, with its call, is self explanatory. The flame-like discharge visible is produced by means of the extreme electrical oscillations which skip through the coil validated, and violently agitate the electrified molecules of the air.

By this indicates, a sturdy affinity is created the various 2 normally detached elements of the surroundings, and that they integrate effects, in spite of the truth that no similarly provision is made for intensifying the chemical movement of the discharge. In the manufacture of nitrogen compounds with the resource of this technique, of route, each feasible way bearing upon the intensity of this movement and the overall performance of the approach might be taken advantage of, and, except, specific preparations may be provided for the fixation of the compounds commonplace, as they'll be typically risky, the nitrogen turning into over again inert after a piece lapse of time. Steam is a easy and effective technique for fixing completely the compounds. The stop end result illustrated makes it functionality to oxidize the atmospheric nitrogen in limitless portions, merely via manner of the use of cheap mechanical strength and clean electric powered powered system.

In this manner many compounds of nitrogen may be synthetic anywhere in the international, at a small cost, and in any desired amount, and via manner of these compounds the soil may be fertilized and its productivity indefinitely improved. An abundance of reasonably-priced and wholesome food, not synthetic, but inclusive of we're familiar with, can also consequently be acquired. This new and inexhaustible supply of meals-deliver may be of incalculable gain to mankind, for it'll quite make a contribution to the growth of the human mass, and as a quit result add immensely to human power. Soon, I choice, the world will see the begin of an organisation which, in time to return returned returned, will, I recall, be in significance next to that of iron.

It is obvious that an entire lot of your paintings has been generations beforehand of its time. I suspect society will ultimately see the actual elegance of your paintings because it will truely be the number one building blocks that permits you to decorate society even in addition, until every other active genius thinker together with you presents dimensions for your art work.

In the interim, permit's leap to the second one trouble you referenced in advance; How to lessen the pressure retarding the human mass and the paintings of telautomatics.

As earlier than said, the strain which retards the onward motion of guy is in factor frictional and in element awful. To illustrate this distinction I can also name, as an instance, lack of know-how, stupidity, and imbecility as some of the genuinely frictional forces, or resistances with none directive tendency. On the alternative hand, visionariness, madness, self-damaging tendency, non secular fanaticism, and so on, are all the forces of a awful man or woman, appearing in unique suggestions. To reduce or absolutely conquer these severa retarding forces, exceedingly extremely good strategies need to be hired. One is aware of, as an example, what a fanatic can also do, and you may take preventive measures, can enlighten, convince, and, probable direct him, turn his vice into one-of-a-kind feature; but one does now not understand, and in no way can realise, what a brute or an imbecile also can do, and one need to deal with him as with a mass, inert, with out mind, let out thru the mad elements. A horrible pressure continuously implies some nice, not every now and then a excessive one, despite the fact that badly directed, which it's miles feasible to show to right benefit; however directionless, frictional pressure involves unavoidable loss. Evidently, then, the number one and cutting-edge way to the above query is: turn all terrible stress within the proper path and decrease all frictional pressure.

There may be absolute confidence that, of all the frictional resistances, the most effective that maximum retards human movement is lack of statistics. Not without motive stated that man of knowledge, Buddha: "Ignorance is the excellent evil in the world." The friction which sooner or later ends up from lack of information, and that is notably extended due to the severa languages and nationalities, may be reduced pleasant with the aid of the spread of information and the unification of the heterogeneous elements of humanity. No try is probably better spent. But however lack of information may additionally have retarded the onward motion of man in times beyond, it's miles sure that, these days, poor forces have emerge as of more significance.

It appears the reference of the "negative pressure" is some different manner to provide an explanation for the act of engaging in war amongst worldwide locations?

Yes, among the ones there may be considered one in all a long way more moments than each different; organized conflict. When we recollect the lots and heaps of people, often the first-rate in mind and frame, the flower of humanity, who're forced to a life of nation of no interest and unproductiveness, the large sums of cash each day required for the upkeep of armies and struggle equipment, representing ever a number of human energy, all of the try uselessly spent in the production of fingers and implements of destruction, the lack of life and the fostering of a barbarous spirit, we are appalled at the inestimable loss to mankind which the lifestyles of these deplorable situations have to include. What can we do to fight this fantastic evil?

Law and order actually require the renovation of organized strain. No community can exist and prosper without inflexible difficulty. Every u . S . Have to be able to guard itself, ought to the need stand up. The situations of to-day aren't the end result of the day gone by, and an intensive alternate cannot be effected to-morrow. If the nations may right away disarm, it is greater than in all likelihood that a rustic of things worse than battle itself might also want to conform with. Universal peace is a cute dream, but not proper away realizable. We have visible presently that even the noble strive of the man or woman invested with the exceptional worldly electricity has been without a doubt without impact. And no wonder, for the set up order of commonplace peace is, inside the period in-between, a bodily impossibility. War is a horrible stress, and can't be have become in a terrific course with out passing via, the intermediate ranges. It is a hassle of creating a wheel, rotating one way, turn inside the opposite course with out slowing it down, preventing it, and rushing it up all all over again the opposite way.

Einstein as quickly as said that he supported the atomic bomb as a way to preventing massive warfare but he later came to regret that guide. What do you agree with you studied of such weaponry?

It has been argued that the perfection of guns of excellent detrimental electricity will save you battle. So I myself concept for a long term, but now I too don't forget this to be a profound mistake. Such tendencies will drastically modify, but no longer arrest it. On the other, I suppose that every new arm that is invented, each new departure that is made in this course, simply invites new information and expertise, engages new strive, gives new incentive, and so first-rate gives a clean impetus to similarly improvement. Think of the invention of gun-powder. Can we conceive of any greater radical departure than became suffering from this innovation? Let us receive as authentic with ourselves living in that period: need to we now not have idea then that conflict became at an prevent, at the equal time as the armor of the knight have end up an item of ridicule, at the same time as bodily energy and ability, which means plenty in advance than, have emerge as of quite little price? Yet gunpowder did not prevent battle: pretty the opposite it acted as a maximum effective incentive.

Nor do I don't forget that war can ever be arrested thru any medical or exceptional improvement, as long as comparable situations to the ones prevailing now exist, due to the fact battle has itself come to be a technology, and due to the reality warfare involves a number of the most sacred sentiments of which guy is capable. In reality, it is doubtful whether or not men who might not be prepared to fight for a immoderate precept is probably right for something the least bit. It isn't always the thoughts which makes guy, neither is it the body; it's far the thoughts and body.

“Our virtues and our failings are inseparable, like strain and be counted extensive variety. When they separate, man is not any extra.”

Another argument, which includes large strain, is frequently made, especially, that warfare want to quick come to be not viable because of the fact the manner of defense are outstripping the method of assault. This is first-rate according with a crucial regulation which can be expressed through the assertion that it is a lot less complicated to destroy than to bring together. This law defines human capacities and human situations. Were the ones such that it might be easier to construct than to damage, man need to go on unresisted, growing and amassing without restriction.

Such conditions aren't of this earth. A being which could do that couldn't be a person: it is probably a god. Defense will continuously have the advantage over attack, however this by myself, it appears to me, can in no manner prevent war. By the use of latest standards of protection we can render harbors impregnable in the direction of assault, however we can't via such approach save you warships assembly in warfare on the excessive sea. And then, if we take a look at this idea to its last development, we're brought about the perception that it is probably better for mankind if assault and protection had been virtually oppositely associated; for if every u . S ., even the smallest, must surround itself with a wall definitely impenetrable, and will defy the rest of the area, a country of things might certainly be added on which would be distinctly awful to human improvement. It is thru abolishing all of the limitations which separate global places and worldwide places that civilization is exquisite furthered.

Again, it is contended with the aid of manner of the use of a few that the appearance of the flying-machine need to bring about state-of-the-art peace. This, too, I believe to be a totally faulty view. The flying-device is absolutely coming, and honestly speedy, however the conditions will continue to be the same as earlier than. In reality, I see no purpose why a ruling electricity, like Great Britain, might not govern the air similarly to the sea. Without wishing to place myself on record as a prophet, I do now not hesitate to say that the next few years will see the repute quo of an "air-power," and its middle may be now not a long way from New York. But, for all that, guys will fight on merrily.

The remarkable development of the war principle might ultimately bring about the transformation of the complete energy of battle into simply ability, explosive power, like that of an electrical condenser. In this shape the struggle-power may be maintained with out try; it would want to be lots smaller in amount, on the same time as incomparably extra powerful.

As regards the safety of a country in competition to distant places invasion, it's miles interesting to look at that it's miles based upon only at the relative, and now not the absolute, range of the humans or rate of the forces, and that, if every u . S . Need to lessen the warfare-stress in the equal ratio, the safety could stay unaltered. An international agreement with the object of lowering to a minimal the conflict-stress which, in view of the triumphing nonetheless imperfect education of the masses, is absolutely important, may additionally, therefore, appear to be the primary rational step to take in the route of diminishing the strain retarding human motion.

Fortunately, the winning situations can not hold indefinitely, for a contemporary-day detail is starting to assert itself. A alternate for the better is coming close to near, and I shall now corporation to show what, in line with my thoughts, may be the number one enlarge toward the set up order of non violent members of the family amongst countries, and by means of manner of what way it's going to in the long run be performed.

Let us skip lower back to the early beginning, while the law of the more potent modified into the only law. The moderate of cause have become not however kindled, and the vulnerable changed into surely at the mercy of the robust. The inclined man or woman then started out to discover ways to defend himself. He made use of a membership, stone, spear, sling, or bow and arrow, and within the path of time, instead of bodily energy, intelligence have become the leader finding out issue inside the warfare. The wild character end up gradually softened through the awakening of noble sentiments, and so, imperceptibly, after a long term of persevered development, we've got come from the brutal combat of the unreasonable animal to what we call the "civilized conflict" of to-day, in which the fighters shake hands, talk in a pleasing way, and smoke cigars within the intermission, however only too organized to engage another time in lethal battle at a sign. Let pessimists say what they prefer, right here is an absolute proof of outstanding and satisfying improve.

But now, what is the subsequent segment in this evolution? Not peace as but, thru any way. The next alternate which must manifestly look at from present day-day inclinations have to be the continuous diminution of the quantity of people engaged in struggle. The device might be considered taken into consideration one in every of mainly superb energy, but only some individuals may be required to carry out it. This evolution will deliver more and more into prominence a device or mechanism with the fewest humans as an element of war, and the really unavoidable cease end result of this could be the abandonment of big, clumsy, slowly transferring, and unmanageable gadgets. Greatest possible velocity and most charge of power-shipping with the useful useful resource of the warfare tool might be the precept object. The loss of lifestyles becomes smaller and smaller, and ultimately, the amount of the humans continuously diminishing, virtually machines will meet in a competition without blood-shed, the worldwide locations being absolutely involved, ambitious spectators. When this happy scenario is decided out, peace can be assured.

But, irrespective of to what diploma of perfection speedy-fireplace guns, high-power cannon, explosive projectiles, torpedo-boats, or one of a kind implements of struggle may be delivered, no matter how destructive they'll be made, that circumstance can in no way be reached through this type of improvement. All such implements require men for his or her operation; guys are essential additives of the gadget. Their object is to kill and to damage. Their power is living in their ability for doing evil. So lengthy as guys meet in war, there may be bloodshed. Bloodshed will ever keep up barbarous passion.

To wreck this fierce spirit, a radical departure have to be made, a totally new precept must be added, some thing that by no means existed before in battle principle a good way to forcibly, always, flip the warfare into an insignificant spectacle, a play, a contest with out lack of blood. To bring forth this end end end result guys need to be allocated with: system have to combat system. But how to accomplish that which seems no longer feasible? The answer is straightforward enough: produce a device able to performing as although it had been a part of a man or woman of mere mechanical contrivance, comprising levers, screws, wheels, clutches, and not anything more, but a system embodying a higher principle, if you need to permit it to carry out its obligations as although it had intelligence, experience, judgment, a mind! This give up is the surrender stop end result of my thoughts and observations that have extended thru absolutely my complete life.

There seems to be, inside the interim, a very famous idea at the functionality for a person to create what they want in truth simply by manner of believing in the pictures in their mind. Do you be given as true with the ones pics assist or thrust back an humans in advance progress?

A long time in the past, as quickly as I changed proper right into a boy, I come to be troubled with a completely unique problem, which appears to had been due to an first rate excitability of the retina. It become the arrival of images which, through their staying energy, marred the vision of real devices and interfered with idea. When a phrase grow to be said to me, the photograph of the item which it superb would possibly appear vividly earlier than my eyes, and commonly it have emerge as now not viable for me to inform whether or not the item I noticed have emerge as real or not. This caused me brilliant pain and anxiety, and I attempted tough to loose myself of the spell. But for a long time I tried in vain, and it come to be no longer, as I without a doubt do not forget, till I become about twelve years vintage that I succeeded for the primary time, by way of an attempt of the need, in banishing an image which supplied itself. My happiness will in no way be as complete as it changed into then, however, alas (as I notion at that element), the antique problem decrease back, and with it my tension. Here it became that the observations to which I refer began out out. I said, specially, that on every occasion the photograph of an item seemed in advance than my eyes I had visible some trouble that jogged my memory of it. In the first instances I notion this to be basically unintentional, however rapid I happy myself that it changed into no longer so. A seen have an effect on, consciously or unconsciously obtained, always preceded the arrival of the image. Gradually the desire arose in me to discover, whenever, what precipitated the images to appear, and the pleasure of this desire speedy have emerge as a need.

The next announcement I made changed into that, certainly as those snap shots placed because of some thing I had visible, so additionally the mind which I conceived were recommended in like manner. Again, I expert the identical choice to locate the picture which brought at the concept, and this look for the genuine visual have an impact on quickly became a 2d nature. My thoughts have turn out to be automated, because it had been, and within the route of years of continued, almost subconscious performance, I obtained the ability of locating every time and, normally, right away the seen impact which started out the idea. Nor is this all.

It modified into no longer prolonged earlier than I modified into aware that still all my moves were caused in the identical manner, and so, searching, staring at, and verifying continuously, 365 days via way of way of yr, I have, by using each concept and each act of mine, tested, and obtain this day by day, to my absolute satisfaction that I am an automaton endowed with energy of movement, which honestly responds to out of doors stimuli beating upon my revel in organs, and thinks and acts and movements therefore. I recollect best one or instances in my existence wherein I modified into no longer capable of find the number one have an effect on which triggered a movement or a concept, or perhaps a dream.

Speaking of desires. I had a dream now not extended within the beyond that every one cell residing creatures on this planet very very own the common function of blood walking via their frame and without this regular go along with the drift, residing cell entities could not exist.

Chapter 3: What Do You Determined Of This Concept And The Way Do You Relate To The Idea That We're On Foot Herbal Robots?

With the reports I related to concerning intellectual imagery, it changed into only herbal that, lengthy inside the past, I conceived the idea of building an automaton which may additionally robotically represent me, and which could reply, as I do myself, however, of course, in a much more primitive manner, to outdoor influences. Such an automaton simply needed to have reason energy, organs for locomotion, directive organs, and one or more sensitive organs so tailored as to be excited via outside stimuli. This system ought to, I reasoned, perform its moves within the manner of a dwelling being, for it would have all of the chief mechanical tendencies or elements of the identical. There changed into though the functionality for growth, propagation, and, mainly, the mind which might be seeking to make the model complete. But growth modified into now not essential in this example, because of the truth that a tool will be synthetic full grown, so to talk. As to the capacity for propagation, it can likewise be not noted of interest, for inside the mechanical model it definitely signified a manner of manufacture.

Whether the automation be of flesh and bone, or of timber and metallic, it mattered little, provided it is able to carry out all the obligations required of it like an intelligent being. To obtain this, it needed to have an detail similar to the thoughts, which might in all likelihood have an impact on the manage of all its actions and operations, and purpose it to act, in any unexpected case that might present itself, with records, cause, judgment, and experience. But this detail I have to with out problems embody in it with the beneficial useful resource of conveying to it my private intelligence, my very personal expertise. So this invention have become advanced, and so a cutting-edge paintings came into lifestyles, for which the choice "telautomatics" has been suggested, because of this the art of controlling the actions and operations of remote automatons.

This precept glaringly modified into applicable to any form of system that moves on land or inside the water or in the air. In utilizing it nearly for the number one time, I determined on a boat (see Fig. 2). A garage battery placed inside it provided the cause electricity. The propeller, pushed by using a motor, represented the locomotive organs. The rudder, managed by the usage of the usage of another motor likewise driven through the usage of the battery, took the location of the directive organs. As to the touchy organ, obviously the primary idea emerge as to make use of a device privy to rays of mild, like a selenium cell, to symbolize the human eye. But upon closer inquiry I decided that, thanks to experimental and different issues, no thoroughly pleasant manage of the automaton is probably tormented by moderate, radiant warmth, hertzian radiations, or via rays in fashionable, this is, disturbances which bypass in right now strains thru vicinity. One of the motives became that any obstacle coming a number of the operator and the a ways flung automaton ought to place it beyond his control.

Another cause was that the touchy device representing the eye may want to should be in a selected role with appreciate to the far flung controlling system, and this necessity can also want to impose extremely good limitations inside the control. Still a few one of a kind and genuinely important motive have emerge as that, within the use of rays, it is probably difficult, if not impossible, to provide to the automaton person functions or developments distinguishing it from special machines of this kind. Evidently the automaton need to reply simplest to an man or woman name, as someone responds to a call. Such concerns led me to finish that the touchy tool of the device have to correspond to the ear in region of the eye of a person, for in this example its moves can be managed irrespective of intervening obstacles, no matter its function relative to the remote controlling tool, and, final, but no longer least, it might continue to be deaf and unresponsive, like a devoted servant, to all calls however that of its maintain close. These necessities made it critical to apply, within the manage of the automaton, in region of moderate or special rays, waves or disturbances which propagate in all instructions thru space, like sound, or which observe a direction of least resistance, but curved. I attained the end result geared toward via an electric powered powered powered circuit placed inside the boat, and modified, or "tuned," exactly to electrical vibrations of the right kind transmitted to it from "electric powered oscillator." This circuit, in responding, but feebly, to the transmitted vibrations, affected magnets and other contrivances, through the medium of which had been managed the movements of the propeller and rudder, and also the operations of numerous first-rate home gadget.

By the easy manner described the information, experience, judgment, the mind, so to speak of the distant operator were embodied in that gadget, which become therefore enabled to transport and to perform all its operations with reason and intelligence. It behaved much like a blindfolded character obeying recommendations obtained via the ear.

The automatons thus far constructed had "borrowed minds," so to talk, as each certainly formed a part of the a ways flung operator who conveyed to it his sensible orders; however this art work is great in the beginning. I cause to expose that, but not possible it may now appear, an automaton can be contrived so you could have its "non-public thoughts," and via this I suggest that it will be succesful, independent of any operator, left absolutely to itself, to perform, in reaction to external affects affecting its sensitive organs, a notable fashion of acts and operations as though it had intelligence. It can be able to study a route laid out or to obey orders given a protracted manner in advance; it will be capable of distinguishing between what it ought and what it ought now not to do, and of making studies or, in any other case said, of recording impressions which will in reality have an effect on its next actions. In reality, I actually have already conceived this type of plan.

Although I advanced this invention some years inside the past and defined it to my web site site visitors very often in my laboratory demonstrations, it turn out to be no longer till a great deal later, long once I had perfected it, that it have end up appeared, even as, in reality enough, it gave rise to a bargain speak and to sensational critiques. But the real significance of this new art changed into no longer grasped thru the majority, nor turned into the amazing pressure of the underlying principle recognized. As nearly as I need to decide from the numerous comments which seemed, the outcomes I had obtained were considered as really now not possible. Even the few who've been disposed to confess the practicability of the discovery observed in it definitely an automobile torpedo, which changed into to be used for the motive of blowing up battleships, with doubtful fulfillment.

The full-size influence have become that I pondered in reality the steering of any such vessel with the useful resource of Hertzian or distinct rays. There are torpedoes advised electrically via wires, and there are method of speaking with out wires, and the above grow to be, of direction, an obvious inference. Had I carried out not whatever more than this, I need to have made a small deliver a boost to virtually. But the paintings I even have advanced does now not ponder merely the alternate of direction of a moving vessel; it offers method of virtually controlling, in every recognize, all the innumerable translatory actions, similarly to the operations of all the internal organs, regardless of how many, of an individualized automaton.

Criticisms to the impact that the manipulate of the automaton is probably interfered with had been made via those who do now not even dream of the super results which can be completed by the use of using the usage of electric powered powered vibrations. The international movements slowly, and new truths are difficult to peer. Certainly, through the use of this principle, an arm for attack further to protection can be furnished, of a destructiveness all the greater because the precept is relevant to submarine and aerial vessels. There are sincerely no rules as to the amount of explosive it can deliver, or as to the gap at which it can strike, and failure is nearly now not feasible. But the stress of this new precept does no longer very well are dwelling in its destructiveness. Its advent delivered into struggle an detail which in no manner existed earlier than stopping-tool with out men as a way of assault and protection. The non-stop improvement in this path should in the end make warfare a trifling contest of machines with out guys and without loss of life situation which would had been impossible without this new departure, and which, in my view, need to be reached as preliminary to everlasting peace. The destiny will each undergo out or disprove those views. My mind on this hassle were located forth with deep conviction, however in a humble spirit.

But to vicinity it virtually, the mounted order of eternal non violent individuals of the own family among international places could most successfully reduce the strain retarding the human mass, and can be the pleasant way to this extremely good human problem. But will the dream of everyday peace ever be decided out? Let us want that it will. When all darkness is probably dissipated through using the mild of technological knowledge, at the same time as all international places shall be merged into one, and patriotism may be equal with religion, while there will be one language, one u . S . A ., one surrender, then the dream becomes reality.

And the 1/3 trouble you furnished. How does that have an impact on the concept of peace and human development? How ought to harnessing the solar's power have this type of dramatic and sturdy effect on human existence on the planet?

Of the 3 feasible solutions of the primary problem of growing human energy, that is thru far the maximum crucial to recollect, not simplest due to its intrinsic importance, but additionally due to its intimate concerning all of the many elements and situations which determine the movement of humanity. In order to hold systematically, it might be essential for me to stay on all of the ones issues that have guided me from the outset in my efforts to attain at an answer, and that have led me, little by little, to the effects I shall now describe.

As a initial study of the problem an analytical studies, which includes I sincerely have made, of the chief forces which decide the onward movement, might be of advantage, specifically in conveying an idea of that hypothetical "speed" which, as described inside the beginning, is a diploma of human strength; but to address this especially right right here, as I must desire, may also want to guide me a long way beyond the scope of the triumphing trouble. Suffice it to u . S . That the ensuing of a number of those forces is constantly in the route of purpose, which therefore, determines, at any time, the direction of human movement. This is to say that each try it's scientifically applied, rational, useful, or practical, ought to be inside the route wherein the mass is moving.

The sensible, rational man, the observer, the man of corporation, he who motives, calculates, or determines in advance, cautiously applies his attempt so that once coming into impact it's going to probable be inside the direction of the motion, making it finally most inexperienced, and in this know-how and potential lies the name of the game of his success. Every new fact decided, every new experience or new element delivered to our records and getting into the domain of purpose, impacts the identical and, consequently, changes the course of motion, which, however, have to always take region alongside the ensuing of all the ones efforts which, at that aspect, we designate as low cost, that is, self-maintaining, beneficial, worthwhile, or sensible. These efforts trouble our each day lifestyles, our necessities and comforts, our paintings and company, and it is the ones which strain guy onward.

But searching at all this busy international about us, on all this complex mass because it every day throbs and moves, what is it however a extremely good clock-artwork pushed with the beneficial aid of a spring? In the morning, whilst we upward push, we can not fail to be conscious that each one the objects approximately us are synthetic by means of device: the water we use is lifted with the useful useful resource of steam-strength; the trains convey our breakfast from far flung localities; the elevators in our living and our place of business constructing, the cars that bring us there, are all pushed by means of manner of the usage of electricity; in all our each day errands, and in our very existence-pursuit, we rely upon it; all the items we see inform us of it; and while we pass decrease lower back to our device-made dwelling at night time time, lest we want to forget about it, all the fabric comforts of our home, our cheering range and lamp, remind us of the manner lots we depend on energy. And at the same time as there is an unintended stoppage of the machinery, whilst the city is snowbound, or the life maintaining movement in any other case in short arrested, we are affrighted to understand how not possible it might be for us to live the existence we stay without purpose energy. Motive energy approach art work. To increase the force accelerating human movement method, consequently, to perform greater artwork.

How may additionally you summarize the 3 factors of developing human strength and the way excellent to suggest societal boom?

So we find out that the 3 feasible answers of the terrific trouble of growing human strength are answered by means of the use of the use of the three terms: food, peace, art work. Many a yr I surely have idea and pondered, out of vicinity myself in speculations and theories, considering man as a mass moved with the resource of a pressure, viewing his inexplicable motion inside the mild of a mechanical one, and utilising the clean thoughts of mechanics to the evaluation of the equal until I arrived at the ones answers, most effective to understand that they had been taught to me in my early formative years. These three terms sound the important element-notes of the Christian religion.

Furthermore, their scientific this means that and reason now smooth to me: food to boom the mass, peace to decrease the retarding pressure, and work to boom the stress accelerating human movement. These are the handiest 3 solutions which might be feasible of that high-quality trouble, and they all have one item, one end, mainly, to boom human energy. When we apprehend this, we can not assist wondering how profoundly clever and medical and the way immensely sensible the Christian faith is, and in what a marked assessment it stands on this appreciate to other religions. It is unmistakably the quit result of practical test and scientific announcement that have prolonged thru the some time, at the same time as first rate religions seem like the outcome of honestly summary reasoning. Work, untiring strive, useful and accumulative, with intervals of relaxation and recuperation aiming at better overall performance, is its chief and ever-ordinary command. Thus we're inspired each with the beneficial useful resource of Christianity and Science to do our utmost closer to developing the overall overall performance of mankind. This maximum critical of human issues I shall now in particular undergo in thoughts.

You have spoken often approximately the supply of human energy. Can you supply an cause for that for us now?

First allow us to invite: Whence comes all the motive power? What is the spring that drives all? We see the ocean rise and fall, the rivers glide, the wind, rain, hail, and snow beat on our domestic home windows, the trains and steamers come and skip; we proper right right here the rattling noise of carriages, the voices from the street; we experience, heady scent, and taste; and we take into account all this. And all this movement, from the surging of the strong ocean to that diffused motion worried in our idea, has however one common motive. All this energy emanates from one single center, one single supply, the solar. The sun is the spring that drives all. The sun keeps all human life and factors all human energy.

Another answer we've got got now located to the above remarkable question: To increase the pressure accelerating human motion manner to turn to the uses of guy greater of the sun's power. We honor and revere those incredible men of bygone times whose names are associated with immortal achievements, who have proved themselves benefactors of humanity the religious reformer collectively along with his clever maxims of life, the reality seeker alongside along with his deep truths, the mathematician alongside along with his approach, the physicist together with his prison hints, find out alongside along with his requirements and secrets and techniques and strategies and techniques wrested from nature, the artist together with his types of the lovely; but who honors him, the best of all,who can inform the call of him who first have end up to apply the sun's power to keep the attempt of a susceptible fellow-creature? That became man's first act of scientific philanthropy, and its outcomes were incalculable.

From the very starting 3 approaches of drawing strength from the solar were open to man. The savage, at the same time as he warmed his frozen limbs at a hearth kindled in a few manner, availed himself of the strength of the solar stored within the burning fabric. When he carried a package of branches to his cave and burned them there, he made use of the solar's saved power transported from one to every different locality. When he set sail to his canoe, he applied the energy of the solar achieved to the environment or the ambient medium. There may be absolute confidence that the primary is the oldest manner. A fireplace, determined by way of risk, taught the savage to understand its beneficial warmth. He then very probably conceived of the idea of carrying the glowing participants to his home. Finally, he found out to apply the stress of a quick contemporary of water or air.

It is feature of current improvement that improvement has been affected within the identical order. The usage of the strength saved in wood or coal, or, usually speaking, gasoline, led to the steam-engine. Next a wonderful stride earlier modified into made in electricity-transportation through the usage of energy, which prison the transfer of electricity from one locality to every other with out transporting the material. But as to using the power of the ambient medium, no radical jump forward has as however been made identified.

The final outcomes of development in the ones 3 instructions are: first, the burning of coal via a chilly device in a battery; second, the green usage of the energy of the ambient medium; and, 1/three the transmission without wires of electrical power to any distance. In whatever manner the ones effects may be arrived at, their sensible software will usually contain an intensive use of iron, and this useful metallic will virtually be an important detail within the similarly development alongside those 3 traces. If we gain burning coal by using a cold method and for this reason gain electric electricity in an efficient and much much less high priced way, we can require in lots of practical makes use of of this power electric powered powered vehicles that is, iron, the same energy in abundance in some unspecified time in the future of our blood and the best supply of that is from the universe itself.

If we are a fulfillment in deriving strength from the ambient medium, we will need, every inside the obtainment and usage of the electricity, device another time, iron. If we realise the transmission of electrical electricity without wires on an industrial scale, we will be pressured to use considerably electric powered generators, all yet again, iron. Whatever we can also furthermore do, iron will probably be the chief technique of achievement inside the near future, probable greater so than in the past. How long its reign will very last is difficult to tell, for even now aluminium is looming up as a threatening competitor. But within the meanwhile, next to presenting new assets of power, it is of the satisfactory importance to growing upgrades inside the manufacture and usage of iron. Great advances are possible within the ones latter guidelines, which, if added about, may fairly increase the useful normal performance of mankind.

In speakme of iron and the need to draw strength from the sun and in the end the universe itself, how will the ones cosmic forces form our destinies?

Every living being is an engine geared to the wheelwork of the universe. Though seemingly affected best with the aid of its without delay surrounding, the sphere of out of doors have an impact on extends to infinite distance. There isn't any constellation or nebula, no solar or planet, in all of the depths of infinite area, no passing wanderer of the starry heavens, that does not exercising a few manage over its future no longer inside the vague and elusive sense of astrology, but inside the rigid and pleasant that means of bodily technological understanding.

More than this may be stated. There is not any element endowed with life from man, who's enslaving the elements, to the humblest creature in this worldwide that doesn't sway it in flip. Whenever movement is born from stress, despite the fact that or now not it's far infinitesimal, the cosmic stability is disenchanted and the conventional motion effects.

Herbert Spencer has interpreted existence as a non-stop adjustment to the surroundings, a definition of this inconceivably complex manifestation pretty in accord with superior scientific belief, but, possibly, not large sufficient to specific our present perspectives. With each leap beforehand inside the investigation of its legal pointers and mysteries our conceptions of nature and its degrees were gaining intensive and breadth.

With that stated, quantum physics has turn out to be a first-rate deliver of scientific research and discovery because it relates to facts existence inside the global or maybe presenting extra statistics about the universe. What do you observed of quantum physics?

In the early stages of highbrow development man changed into aware about but a small part of the macrocosm. He knew not anything of the wonders of the microscopic global, of the molecules composing it, of the atoms making up the molecules and of the dwindling small international of electrons in the atoms. To him existence turn out to be synonymous with voluntary movement and movement. A plant did no longer suggest to him what it does to us that it lives and feels, fights for its lifestyles, that it suffers and enjoys. Not only have we located this to be proper, however we have ascertained that even depend referred to as inorganic, believed to be useless, responds to irritants and offers unmistakable evidence of the presence of a living precept inner.

Thus, everything that exists, herbal or inorganic, active or inert, is susceptible to stimulus from the out of doors. There is not any hollow amongst, no harm of continuity, no specific and distinguishing vital agent. The identical law governs all don't forget, all of the universe is alive. The momentous question of Spencer, "What is it that reasons inorganic depend to run into natural workplace paintings!" has been answered. It is the sun's warmth and light. Wherever they are there's lifestyles. Only inside the boundless wastes of interstellar region, in the eternal darkness and bloodless, is animation suspended, and, probable, at a temperature of absolute zero all depend might also die.

Chapter 4: Does That Suggest You Do Not Forget Guy A Device?

This realistic issue of the perceptible universe, as a clockwork wound up and walking down, meting out with the need of a hyper mechanical vital principle, want not be in discord with our spiritual and inventive aspirations those undefinable and notable efforts through which the human mind endeavors to free itself from fabric bonds. On the opposite, the better statistics of nature, the eye that our expertise is actual, can handiest be all the extra elevating and scary.

It changed into Descartes, the incredible French reality seeker, who in the seventeenth century, laid the primary basis to the mechanistic principle of existence, now not a touch assisted thru the use of Harvey's epochal discovery of blood flow into. He held that animals had been really automata with out interest and identified that man, despite the fact that possessed of a higher and unique remarkable, is incapable of motion apart from those function of a device. He furthermore made the primary attempt to offer an explanation for the bodily mechanism of reminiscence. But in this time many competencies of the human body have been now not as but understood, and on this apprehend a number of his assumptions were inaccurate.

Great strides have thinking about the truth that been made in the artwork of anatomy, physiology and all branches of science, and the workings of the individual-device are honestly flawlessly clear. Yet the very fewest among us are capable of trace their movements to primary out of doors reasons. Lt is crucial to the arguments I shall increase to preserve in thoughts the primary records which I without a doubt have myself mounted in years of near reasoning and remark and which can be summed up as follows:

1. The guy or women is a self-propelled automaton completely beneath the manage of outdoor affects and best able to studying from the confines of his herbal environment. Of the world he have become born into. Willful and predetermined despite the fact that they appear, his actions are ruled not from inner, but from with out. He is type of a go together with the drift tossed approximately by way of the waves of a turbulent sea.

2. There is no reminiscence or retentive faculty based totally on lasting impressions. What we designate as memory is however accelerated responsiveness to repeated stimuli.

3. It isn't actual, as Descartes taught, that the mind is an accumulator. There isn't any everlasting record inside the thoughts, there may be no stored data. Knowledge is some thing corresponding to an echo that wishes a disturbance to be referred to as into being.

four. All knowledge or shape idea is evoked through the medium of the eye, either in reaction to disturbances right now obtained at the retina or to their fainter secondary outcomes and reverberations. Other feel organs can nice name forth feelings which have no fact of life and of which no idea can be usual

5. Contrary to the most vital guiding principle of Cartesian philosophy that the perceptions of the thoughts are illusionary, the eye transmits to it the real and correct likeness of outdoor things. This is because mild propagates in instantly strains and the image solid at the retina is an real reproduction of the outside shape and one which, way to the mechanism of the optic nerve, can not be distorted in the transmission to the mind. What is greater, the technique need to be reversible, that is to say, a form brought to recognition can, through reflex motion, reproduce the actual picture at the retina virtually as an echo can

reproduce the proper disturbance If this view is borne out thru test an great revolution in all human participants of the own family and departments of hobby can be the impact.

Since power in no way dies and we're smart "residing" electricity how do natural forces have an impact on us then? I imply, do you consider in a better strength? In God?

Accepting all this as real allow us to don't forget a number of the forces and influences which act in this form of splendidly complex automated engine with organs inconceivably touchy and sensitive, as it's miles carried with the useful resource of the spinning terrestrial globe in lightning flight via vicinity. For the sake of simplicity we may also moreover count on that the earth's axis is perpendicular to the ecliptic and that the human automaton is at the equator. Let his weight be 160 pounds then, on the rotational pace of about 1,520 ft in step with 2d with which he's whirled round, the mechanical power stored in his body may be nearly five,780,000 foot kilos, which is about the strength of a hundred-pound cannon ball.

This momentum is steady as well as upward centrifugal push, amounting to approximately fifty-5 hundredth of a pound, and each will probable be without marked have an impact on on his lifestyles abilities. The sun, having a mass 332,000 times that of the earth, however being 23,000 instances farther, will enchantment to the automaton with a stress of approximately one-10th of one pound, alternately developing and diminishing his normal weight by the use of that quantity

Though no longer conscious of those periodic changes, he is certainly suffering from them.

The earth in its rotation throughout the solar consists of him with the prodigious pace of 19 miles consistent with 2nd and the mechanical strength imparted to him is over 25,a hundred and sixty,000,000 foot kilos. The biggest gun ever made in Germany hurls a projectile weighing one ton with a muzzle speed of 3,700 toes regular with second, the energy being 429,000,000 foot kilos. Hence the momentum of the automaton's body is almost sixty times more. It might be enough to increase 762,400 horse-strength for one minute, and if the motion were all of sudden arrested the frame is probably right away exploded with a stress sufficient to keep a projectile weighing over sixty lots to a distance of twenty-eight miles.

This awesome electricity is, but, now not regular, but varies with the position of the automaton with reference to the solar. The circumference of the earth has a pace of one,520 feet constant with 2d, that is each brought to or subtracted from the translatory tempo of 19 miles through location. Owing to this the energy will range from twelve to twelve hours with the useful resource of an quantity approximately identical to as a minimum one,533,000,000 foot kilos, because of this that electricity streams in some unknown way into and out of the body of the automaton on the charge of about sixty-4 horse-electricity.

But this isn't always all. The complete solar tool is urged closer to the a protracted manner flung constellation Hercules at a tempo which a few estimate at a few twenty miles in keeping with second and due to this there need to be comparable annual modifications inside the flux of power, which also can moreover reap the appalling decide of over 100 billion foot kilos. All these numerous and in number one phrases mechanical results are rendered more complex via the inclination of the orbital planes and masses of different permanent or casual mass actions.

This automaton is, however described, subjected to unique forces and impacts. His frame is at the electrical capability of billion volts, which fluctuates violently and often. The complete earth is alive with electric powered vibrations in which he's taking issue. The surroundings crushes him with a stress of from sixteen to 20 plenty, in line with barometric situation. He receives the strength of the solar's rays in diverse periods at a median price of about 40 foot kilos steady with 2d, and is subjected to periodic bombardment of the sun's debris, which skip via his frame as even though it have been tissue paper. The air is hire with sounds which beat on his eardrums, and he is shaken thru way of the unceasing tremors of the earth's crust. He is exposed to superb temperature changes, to rain and wind.

What surprise then that during this sort of terrible turmoil, in which forged iron lifestyles must seem now not feasible, this sensitive human engine ought to act in an exquisite way? If all automata were in each understand alike they might react in exactly the identical manner, but this is not the case. There is concordance in reaction to the ones disturbances which can be most usually repeated, now not to all. It is pretty smooth to provide electric structures which, whilst subjected to the identical effect, will behave in most effective the opposite way.

So moreover humans, and what's real of humans moreover holds unique for his or her massive aggregations. We all sleep periodically. This is not an vital physiological necessity any extra than stoppage at durations is a call for for an engine. It is clearly a scenario progressively imposed upon us via the diurnal revolution of the globe, and that is one of the many evidences of the truth of mechanistic precept. We take a look at a rhythm or ebb and tide, in thoughts and reviews, in economic and political actions, in each department of our intellectual hobby. So fantastic, because the strength of every residing mechanism is completely relying at the deliver mechanism, consequently there want to be One before there can be many.

But you communicate of friction limiting the increase of guy and this same strength can be harnessed to enhance lifestyles for man. Can you deliver an cause for that?

It most effective suggests that in all this a physical machine of mass inertia is concerned which gives a further putting proof. If we get keep of the concept as a essential truth and, furthermore, growth the boundaries of our experience perceptions past those internal which we emerge as aware about the outside impressions, then all of the states in human lifestyles, however unusual, can be plausibly described. A few examples may be given in example.

The eye responds pleasant to mild vibrations via a sure as an opportunity slim variety, however the limits aren't sharply described. It is likewise affected by vibrations past, fine in lesser degree. A individual might also moreover hence turn out to be privy to the presence of each distinct in darkness, or via intervening obstacles, and people laboring beneath illusions ascribe this to telepathy. Such transmission of notion is absurdly now not viable.

The skilled observer notes without problem that those phenomena are because of concept or twist of fate. The same can be said of oral impressions, to which musical and imitative human beings are in particular willing. A person proudly owning those trends will regularly respond to mechanical shocks or vibrations which is probably inaudible.

To mention any other instance of quick-term interest reference may be made to dancing, which includes certain harmonious muscular contractions and contortions of the body in reaction to a rhythm. How they grow to be in style genuinely now, can be satisfactorily described via supposing the life of some new periodic disturbances inside the environment, which might be transmitted via the air or the ground and can be of mechanical, electrical or particular man or woman.

Chapter 5: Comparable Excellent States Of Society

Though it may seem so, a battle can never be because of arbitrary acts of man.

It is constantly the greater or less direct quit end result of cosmic disturbance in which the solar is chiefly concerned.

In many international conflicts of historical record which have been caused through famine, pestilence or terrestrial catastrophes the direct dependence of the solar is unmistakable. But in maximum times the underlying primary reasons are numerous and tough to hint.

In the triumphing conflict it is probably mainly difficult to reveal that the reputedly willful acts of a few human beings were no longer causative. Be it so, the mechanistic precept, being founded on reality confirmed in normal enjoy, virtually precludes the possibility of this shape of kingdom being some thing however the inevitable impact of cosmic disturbance.

The question manifestly gives itself as to whether or no longer there may be some intimate relation amongst wars and terrestrial upheavals. The latter are of determined have an impact on on temperament and disposition, and may at instances be instrumental in accelerating the conflict but other than this there appears to be no mutual dependence, despite the fact that each can be due to the identical number one motive.

What may be asserted with ideal self belief is that the earth can be thrown into convulsions through mechanical results along with are produced in cutting-edge conflict. This announcement can be startling, but it admits of a easy clarification.

Earthquakes are basically due to causes: subterranean explosions or structural changes. The former are known as volcanic, incorporate large energy and are tough to begin. The latter are named tectonic; their electricity is comparatively insignificant and they will be as a result of the slightest marvel or tremor. The common slides within the Culebra are displacements of this kind.

Are you suggesting that warfare, natural calamities, together with earthquakes, and energy are interrelated?

Theoretically, it may be said that one may think approximately a tectonic earthquake and motive it to arise due to the idea, for virtually preceding the discharge the mass may be within the maximum touchy stability. There is a famous mistakes in regard to the electricity of such displacements. In a case nowadays said as quite exceptional, extending as it did over a extensive territory, the power became anticipated at sixty 5,000,000,000,000 foot tons. Assuming even that the entire art work modified into finished in a unmarried minute it might only be equal to that of

7,500,000 horse-electricity in the course of three hundred and sixty five days, which appears tons, but is little for a terrestrial upheaval. The electricity of the sun's rays falling at the equal region is one thousand times greater.

The explosions of mines, torpedoes, mortars and weapons increase reactive forces on the floor which is probably measured in loads or perhaps masses of thousands and make themselves felt everywhere in the globe. Their effect, but, can be specially magnified thru resonance. The earth is a sphere of a anxiety barely more than that of metallic and vibrates as quick as in approximately one hour and forty-9 mins.

If, as may be viable, the concussions arise to be nicely timed their mixed movement ought to start tectonic changes in any part of the earth, and the Italian calamity can also consequently were the surrender result of explosions in France. That guy can produce such terrestrial convulsions is past any doubt, and the time may be close to at the same time as it'll possibly be completed for features correct or apt.

Whoever wants to get a actual appreciation of the greatness of our age must examine the records of electrical development. There he is going to discover a story extra extraordinary than any story from Arabian Nights. It starts offevolved offevolved lengthy earlier than the Christian technology even as Thales, Theophrastus and Pliny tell of the magic homes of electrons, the valuable substance we name amber that got here from the herbal tears of the Heliades, sisters of Phaeton, the unfortunate youngsters who attempted to run the blazing chariot of Phoebus and almost burned up the earth. It have end up however herbal for the tremendous creativeness of the Greeks to ascribe the mysterious manifestations to a hyperphysical motive, to endow the amber with life and with a soul.

Whether this became real perception or genuinely poetic interpretation remains a query. When at this very day maximum of the maximum enlightened people assume that the pearl is alive, that it grows more lustrous and first-rate in the warmness contact of the human body. So too, it's far the opinion of fellows of technological knowledge that a crystal is a dwelling being and this view is being prolonged to encompass the entire physical universe because Prof. Jagadis Chunder Bose has mounted, in a chain of first-rate experiments, that inanimate depend responds to stimuli as plant fiber and animal tissue.

The superstitious perception of the ancients, if it existed the least bit, can consequently no longer be taken as a dependable proof in their lack of knowledge, however simply how a good deal they knew about electricity can simplest be conjectured. A curious truth is that the ray or torpedo fish changed into used by them in electro-treatment. Some vintage coins display twin stars, or sparks, along facet is probably produced by manner of a galvanic battery. The statistics, even though scanty, are of a nature to fill us with conviction that a few initiated, as a minimum, had a deeper know-how of amber-phenomena. To point out one, Moses turn out to be in reality a sensible and skillful electrician an extended way in advance of his time. The Bible describes exactly and minutely arrangements constituting a system wherein electricity have become generated with the resource of friction of air towards silk curtains and stored in a field constructed like a condenser. It can be very viable to count on that the sons of Aaron were killed by a excessive anxiety discharge and that the vestal fires of the Romans had been electric. The belt force need to had been recognised to engineers of that epoch and it's miles hard to look how the adequate evolution of static strength may also need to have escaped their word. Under favorable atmospheric conditions a belt can be converted proper into a dynamic generator capable of generating many placing movements. I really have lighted incandescent lamps, operated motors and executed numerous extraordinary further thrilling experiments with electricity drawn from belts and saved in tin cans.

That many facts in regard to the diffused force were recounted to the philosophers of vintage may be appropriately concluded, the wonder is, why thousand years elapsed in advance than [William Gilbert] in 1600 posted his well-known paintings, the number one scientific treatise on power and magnetism. To an amount this long length of unproductiveness may be defined. Learning turned into the privilege of a few and all facts end up jealously guarded. Communication have become hard and sluggish and a mutual knowledge among extensively separated investigators tough to reach. Then yet again, men of these instances had no concept of the realistic, they lived and fought for summary concepts, creeds, traditions and ideals.

Humanity did now not trade masses in Gilbert's time however his smooth teachings had a telling effect at the minds of the located. Friction machines were produced in rapid succession and experiments and observations extended. Gradually worry and superstition gave manner to scientific in-sight and in 1745 the region changed into thrilled with the information that Kleist and Leyden had succeeded in imprisoning the uncanny agent in a phial from which it escaped with an angry snap and unfavorable pressure. This changed into the shipping of the condenser, probably the maximum mind-blowing electrical tool ever invented.

Two fantastic leaps had been made in the succeeding 40 years. One modified into even as Franklin established the identification among the mild soul of amber and the awe-inspiring belt of Jupiter; the alternative at the same time as Galvany and Volta introduced out the contact and chemical battery, from which the magic fluid can be drawn in endless quantities. The succeeding forty years bore despite the fact that extra fruit. Oersted made a significant enhance in deflecting a magnetic needle by way of manner of an electric powered powered powered cutting-edge, Arago produced the electro-magnet, Seebeck the thermo-pile and in 1831, because the crowning fulfillment of all, Faraday introduced that he had obtained energy from a magnet, therefore discovering the principle of that first rate engine the dynamo, and inaugurated a modern-day generation each in scientific studies and realistic utility.

From that issue on innovations of inestimable fee have observed every other at a bewildering rate. The telegraph, smartphone, phonograph and incandescent lamp, the induction motor, oscillatory transformer, Roentgen ray, Radium, wi-fi and severa exclusive contemporary advances have been made and all situations of existence eighty-four years that have for the reason that elapsed, the diffused dealers residing within the dwelling amber and lodestone were converted into cyclopean forces turning the wheels of human improvement with ever growing tempo. This, in short, is the fairy story of electricity from Thales to the cutting-edge-day. The not possible has came about, the wildest dreams had been passed and the astounded global is asking: What is coming subsequent?

Chapter 6: What Was Nikola Tesla Like?

Nikola Tesla changed right into a Serbian-American innovator, electric engineer, mechanical engineer, and futurist who helped establish the cutting-edge rotating present (AC) electric power delivery tool.

Tesla, who changed into born and raised inside the Austrian Empire, studied engineering and physics without getting a grade in the 1870s and acquired useful revel in in the early Eighteen Eighties walking in cellular phone and at Continental Edison within the nascent electric powered power corporation. He relocated to america within the one year 1884 and ended up being a naturalized person. Before starting up on his very own, he labored for a fast time on the Edison Device Works in N.Y. City. Tesla advanced labs and industrial organization in New York to boom a number of electric and mechanical devices with the useful aid of partners to fund and promote his thoughts. His rotating present (AC) induction motor and associated polyphase AC trends have been prison via Westinghouse Electric inside the year 1888 and were the muse of the polyphase gadget that business enterprise consequently marketed.

Let's have a look at more approximately this nutty professor, this obsessed genius who changed into way in advance of his time.

Tesla tampered with mechanical oscillators/generators, electric powered discharge tubes, and early X-ray imaging so that it will produce patentable and valuable improvements. He additionally created a wireless-managed boat, which became just one of the first to be proven. Tesla rose to prominence as a developer, displaying his achievements at his laboratory to celebs and wealthy clients, and become extensively identified for his theatrics at public talks. In his excessive-voltage, immoderate-frequency electricity exams in New York and Colorado Springs in the route of the Eighteen 1990s, Tesla went after his thoughts for cordless lights and spherical the arena cordless electric powered energy glide. He made declarations within the three hundred and sixty five days 1893 on the possibilities of cordless communication alongside with his devices. Tesla tried to apply these mind in his incomplete Wardenclyffe Tower mission, a worldwide cordless communique and energy transmitter, but he ran out of coins earlier than completing it.

Tesla tampered with a succession of enhancements after Wardenclyffe in the 1910s and 1920s, with numerous stages of success. Tesla lived in positive New York resorts after spending the majority of his coins and leaving unsettled prices. In January 1943, he passed away in New York City. Following his lack of lifestyles, Tesla's improvement diminished into obscurity until 1960, when the General Conference on Weights and Steps brought the tesla the SI device of magnetic flux denseness in his honor. Since the 1990s, there was a renaissance in public interest in Tesla.

From 1888 till approximately 1926, Tesla end up six ft 2 inches (1.88 m) tall and weighed 142 pounds (sixty four kg), with almost no weight model. "Practically the first-class, almost the thinnest, and maximum probably the maximum extreme man this is going to Delmonico's frequently," paper editor Arthur Brisbane stated of him. In NY City, he emerge as an fashionable, advanced determine who have come to be fastidious in his grooming, outfit, and each day sports, an photo he preserved to boost his commercial corporation contacts. He moreover had colourful eyes, "highly super arms," and "extremely large" thumbs, in step with reports.

Eidetic Memory

Tesla have become a starved reader who remembered complete volumes and come to be said to have a photographic reminiscence.

He changed into a polyglot, speakme Serbo-Croatian, Czech, English, French, German, Hungarian, Italian, and Latin, to call some languages.

In his autobiography, Tesla tremendous that he had a few specific second of motivation. Tesla turn out to be affected with contamination in some unspecified time in the future of his kids. He had an uncommon situation in which he ought to see blinding flashes of mild within the the front of his eyes, usually joined by means of pics. The visions had been frequently related to a term or idea he had located; at excellent instances, they supplied the solution to a specific hassle he had treated. He have to recall a product in realistic detail in fact via way of way of listening to its name. Before carrying right away to the building degree, Tesla observed a development in his head with outstanding detail, incorporating all measurements, a way called image questioning. He did no longer typically draw with the resource of hand; as an opportunity, he trusted recollection. Tesla had routine flashbacks to occasions that had taken region in advance in his life due to the fact he have come to be a younger baby.

His Relationships

Tesla have emerge as an prolonged-lasting bachelor who as soon as cited that his chastity helped his scientific skills drastically.

He has previously referred to that he absolutely believes he may in no manner ever gain enough for a lady, believing ladies to be super in each manner. Later on in existence, he began to trade his mind at the same time as he observed out that women have been looking to surpass men and turn out to be being greater powerful. Tesla grow to be outraged via using this "new woman," believing that by using manner of manner of attempting to find to be powerful, women were dropping their womanhood. On the tenth of August 1924, he specified in an interview with the Galveston Daily News, "In vicinity of the clean-voiced, reputable gentlewoman, has come the female who simply believes that the crucial issue to her fulfillment in existence is to appearance as much like a real guy as viable-- in get dressed, voice, and movements, in sports sports activities and accomplishments of a sizeable range ... I'm disenchanted thru way of girls's propensity to push guys apart, converting the vintage spirit of collaboration with him in all elements of life." Tesla decided on to in no way ever chase after or take part in any diagnosed relationships, instead locating all of the stimulus he desired in his machine, at the identical time as later telling a press reporter that he at instances felt he had made undue a sacrifice to his career with the resource of no longer bridal ceremony.

Tesla became an introvert who decided on to be on my own alongside collectively along with his artwork.

When he did take part in social sports activities, even though, loads of people talked fairly of Tesla and applauded him. He had "outstanding sweet taste, sincerity, modesty, beauty, kindness, and strength," normal with Robert Underwood Johnson. "His adorable smile and splendour of disposition normally signified the gentlemanly traits that have been so ingrained in his character," Dorothy Skerrit, his secretary, wrote. "Hardly ever did one meet a researcher or engineer who grow to be also a poet, a theorist, an appreciator of stunning track, a linguist, and a lover of food and drinks," wrote Tesla's top buddy Julian Hawthorne.

Francis Marion Crawford, Robert Underwood Johnson, Stanford White, Fritz Lowenstein, George Scherff, and Kenneth Swezey were all suitable buddies of Tesla's.

Tesla ended up being a pal of Mark Twain in his later years, and the 2 spent a whole lot of time collectively in his laboratory and in other locations. Tesla's induction motor improvement changed into known as "the most important patent for the reason that phone" thru Mark Twain. Tesla met Indian Hindu monk Swami Vivekananda at a reception hosted via starlet Sarah Bernhardt in the three hundred and sixty 5 days 1896. Tesla stated that he want to mathematically show the relationship among depend and strength, which Vivekananda felt ought to supply Vedantic cosmology a clinical grounding. Tesla befriended George Sylvester Viereck, a poet, author, mystic, and in the end Nazi propagandist, inside the late 1920s. Tesla sometimes went to Viereck and his associate's supper events.

Tesla is probably immoderate every now and then, and he overtly unique his ridicule for overweight humans, like at the same time as he fired a secretary because of her weight.

Tesla become very brief to slam garments, and on numerous occasions, he advocated a secondary to go home and change.

Tesla gave The New York Times the unmarried essential view approximately Thomas Edison even as he handed away within the 365 days 1931, buried in a prolonged insurance of Edison's lifestyles:

"He did no longer have a enjoyment hobby, did no longer care about amusement of any kind, and resided in whole disrespect of the maximum popular fitness legal guidelines ... His method mishandled to the immoderate, considering a massive vicinity needed to be surpassed through to accumulate whatever besides good fortune stepped in, and I became almost a sorry witness to his deeds in the starting, information that a chunk concept and estimation may want to have stored him ninety% of the difficult work. He, as a substitute, hated ebook learning and math know-how, relying completely on his innovator's instinct and beneficial American perceptiveness.

Tesla's Sleeping Patterns

Tesla said that he in no way ever slept for extra than 2 hours at a time.

He did, even though, admit to "napping" now and again to "recharge his batteries."

Tesla advanced a sturdy capabilities for billiards, chess, and card-playing in the direction of his 2d three hundred and sixty five days of research studies at Graz, spending up to two days at a sport desk at a time.

At one issue in his lab, Tesla labored for 80 four hours immediately with out preventing.

Tesla seldom slept, steady with Kenneth Swezey, a reporter whom Tesla had befriended. Swezey recollects being awoke at three a.M. By approach of a call from

Tesla: "I turn out to be sound asleep off in my room like a body ... The ringing of the phone jolted me wide awake ... Tesla talked animatedly, with stops in quick, whilst he ... Worked out an problem, evaluating one hypothesis to some other, making comments; and while he felt he had come to a decision, he rapid closed the cellular telephone."

Tesla's Work Set up

Tesla worked from 9:00 a.M. Until 6:00 p.M. Or later each day, with dinner at eight:10 p.M. At Delmonico's and later the Waldorf-Astoria Hotel. Tesla then located his dinner order with the headwaiter, who is probably the only everybody that could serve him. "At eight:00 p.M., the lunch had to be prepared ... He ate on my own, save for the few situations at the equal time as he had to feed a fixed to meet his social dedications. Tesla lower back to paintings, typically till 3 a.M."

Tesla walked amongst eight and ten miles (13 and 16 kilometers) a day for workout. Every night time, he curled his ft a hundred times for every foot, saying that it caused his brain cells.

Tesla noted that he did no longer accept as actual with in telepathy in an interview with paper editor Arthur Brisbane, saying, ""Suppose I made up my thoughts to murder you," he introduced, "you can recognise it in a second." That's incredible, isn't always it? What system does the mind use to gather all of this?" Tesla particular inside the actual equal interview that he without a doubt believes all number one laws is probably minimized to at the least one.

In his final years, Tesla ended up being a vegetarian, enduring on milk, honey, bread, and some veggie juice he preferred.

Chapter 7: His Beliefs And Point Of View On Life

Tesla refuted the idea that atoms are made up of littler subatomic debris, announcing that an electron couldn't produce an electrical rate. Tesla felt that if electrons existed the least bit, they have been a 4th us of a of do not forget or "sub-atom" that could handiest exist in a speculative vacuum and had actually nothing to do with electrical strength. He had religion within the nineteenth-century idea of an all-pervasive ether that added electric powered power.

Tesla modified into usually opposed to theories which include the change of preserve in thoughts into energy. He furthermore slammed Einstein's concept of relativity, pronouncing:

Area, in my mind-set, cannot be bent because it does no longer have attributes. It is nearly as despite the fact that God is endowed with tendencies. He does not have any, only our very very own tendencies. When dealing with rely that fills location, we're capable of most effective point out trends. To endorse that area eventually finally ends up being curved inside the presence of big our our bodies is to say that honestly not anything can run on a few factor. I, for one, cannot take shipping of as actual with the sort of angle.

Tesla stated to have completed a "colourful concept of gravity" that" [would] positioned an forestall to nugatory theories and wrong ideas, like that of curved area" in a letter written in the three hundred and sixty five days 1937 on the age of 80 one. He stated that the hypothesis have been "worked out in all specifics" and that he wanted to offer it to the area fast.

Tesla's biographers generally concur that, further to his technical competencies, he became a philosophical humanist. Nonetheless, Tesla, like many others of his duration, ended up being an endorse of an enforced selective breeding shape of eugenics.

Tesla simply believed that mankind's "pity" had obstructed of nature's "callous operations." He argued for eugenics, no matter the reality that he did no longer consider in a "hold close race" or in the intrinsic supremacy of a single character over a few other. He mentioned in a 1937 interview, "

"Man's new enjoy of pity started out to disrupt nature's callous operations. The handiest approach appropriate with our standards of society and race is to stop the breeding of the now not worth thru sanitation and the intentional help of the breeding impulse ... The pattern amongst eugenists is that we need to make marriage greater tough. Definitely no person who isn't a pinnacle discern wants to be allowed to offer youngsters.

Tesla mentioned the faults of ladies's social subordination and their undertaking for intercourse equality within the year 1926, searching out that "Queen Bees" will manage mankind's destiny. In the future, he said, ladies could in all likelihood surpass guys because the dominant sex.

In a printed put up entitled "Science and Discovery are the Great Forces Which Will Cause the Consummation of the War" (20 December 1914), Tesla made forecasts about the proper issues of a placed up-World War I environment. Tesla simply believed that the League of Nations wasn't a resolution for the instances and issues.

Tesla grew up in an Orthodox Christian home. Later in lifestyles, he stated that he wasn't a "believer in the orthodox experience," that he unfavorable religious fanaticism, and that "Buddhism and Christendom are the first-rate faiths every in huge form of disciples and in importance." He moreover stated that "To me, the large universe is best a exceptional device that in no way ever started out out gift and never ever will end," and that "what we name 'soul' or 'spirit,' is certainly not anything more than the amount of the performances of the body. When this functioning stops, the 'soul' or the 'spirit' stops moreover."

Chapter 8: His Youth And Years At Edison

Nikola Tesla modified into born on July the 10th, in the 12 months 1856 as an ethnic Serb inside the town of Smiljan, within the Armed stress Frontier, inside the Austrian Empire (contemporary-day-day Croatia). Milutin Tesla (1819-- 1879), his dad, modified into an Eastern Orthodox Church priest.

Tesla's mother, uka Mandi (1822-- 1892), had a present for making domestic craft devices and mechanical domestic home equipment, and the capability to consider Serbian mythical literature, and her dad changed into furthermore an Eastern Orthodox Church priest. Uka had in no manner ever participated in an dependable college. Tesla ascribed his mother's DNA and impact for his eidetic reminiscence and progressive capability. Tesla's forefathers came from western Serbia, close to Montenegro.

Tesla modified into the 4th youngster in a own family of 5. Milka, Angelina, and Marica were his three sisters, and he had a senior brother referred to as Dane, who handed away in a horseback the use of mishap while Tesla become 5 years of age. Tesla studied German, math, and non secular belief at a number one college in Smiljan inside the year 1861. The Tesla family moved to Gospi inside the 12 months 1862, in which Tesla's dad labored as a parish priest. Nikola completed essential college and after that proceeded to intermediate faculty. Tesla transferred to Karlovac inside the 365 days 1870 to take part inside the Higher Real Gym, in which the commands were taught in German, as held real inside the course of the Austro-Hungarian Armed Force Frontier.

Tesla consequently said that his physics professor ignited his interest in electrical strength displays.

These show monitors of this "mystery phenomenon" made Tesla need to "realize greater approximately this first-rate power," he said. Tesla's functionality to finish essential calculus in his thoughts led his instructors to simply accept as right with he changed into cheating. He finished inside the year 1873 after finishing a four-year time period in 3 years.

Tesla went lower again to Smiljan after graduation, but all of sudden advanced cholera, modified into bedridden for 9 months, and came close to demise some times. Tesla's dad (who had to begin with without a doubt preferred him to turn out to be being a clergyman) assured him that if he recuperated from his contamination, he ought to deliver him to the principle engineering college.

Tesla prevented conscription into the Austro-Hungarian Army the following 3 hundred and sixty 5 days in Smiljan through escaping southeast of Lika to Tomingaj, close to Graac. He went there impersonated a hunter and checked out the mountains.

Tesla stated that his contact with nature made him bodily and intellectually greater powerful. While at Tomingaj, he studied an entire lot of books and later said that Mark Twain's works had astonishingly helped him recover from his preceding health trouble.

On an Armed strain Frontier researchership, he registered on the Imperial-Royal Technical College in Graz inside the 365 days 1875. Tesla stated in his autobiography that he labored very difficult and were given the satisfactory grades possible, passing nine tests (nearly times as many as wanted , and getting a letter of gratitude from the dean of the technical college to his dad, citing, "Your toddler is a superstar of first rank." Tesla mentioned his entertainment with Teacher Jakob Pöschl's entire lectures on electrical strength and the way he made thoughts to beautify the layout of an electrical motor the professor became showing at Graz.

Though, thru using his third three hundred and sixty five days, he grow to be failing training and left, leaving Graz inside the month of December 1878. Tesla could have been ejected for gaming and womanizing, regular with one biographer.

Following his departure from Graz, Tesla lessen all ties alongside collectively together with his family.

His right friends presumed he drowned in the near Mur River , despite the fact that he changed into found with the useful resource of a first rate pal in Maribor, Slovenia, in which he become strolling as a draftsman for sixty florins a month, and can have persisted thru Zagreb to a piece settlement at the Adriatic Sea's coast.

Milutin in the end located him in the month of March 1879 and made an attempt to influence him to return lower back domestic and keep his college in Prague. Later on that month, Tesla became deported from Gospi for not having a legitimate residency license. Tesla's dad surpassed away a month later, on April seventeenth, 1879, on the age of 60, from an unknown sickness (a few assets say it can have been a stroke). Tesla taught a sizeable elegance of children in his old fashioned in Gospi for the the rest of the 12 months.

2 of Tesla's uncles pooled their rate variety in the month of January 1880 to allow him depart Gospi for Prague, in which he prepared to test. He got here an extended manner too past because of sign up at Charles-Ferdinand University; he had never ever studied Greek, that have end up an crucial route; and he became illiterate in Czech, which was moreover a call for. Tesla did, in spite of the truth that, participate in philosophy lectures as an auditor on the college, however he failed to earn grades for the guides.

Tivadar Puskás landed Tesla a modern-day undertaking with the Continental Edison Company in Paris inside the yr 1882.

Tesla commenced out out his career in a trendy market, installing indoor incandescent lighting in big electric powered power energies throughout the state. Tesla worked for the Société Electrique Edison, part of the corporation based totally inside the Paris community of Ivry-sur-Seine that supervised of developing the lighting device. He have been given plenty of real electric engineering enjoy there. His first rate engineering and physics know-how stuck manage's attention, and he end up brief developing and producing boosted variations of creating eager beavers and cars. They furthermore despatched him to tackle engineering problems at different Edison energies now under constructing and construction in France and Germany.

Making the relocate to the united states

In the only year 1884, Edison supervisor Charles Batchelor, who have been in price of the Paris setup, emerge as moved to the Edison Device Works, a manufacturing subsidiary in New York City, and asked for that Tesla get preserve of the U.S. Too.

Tesla traveled within the month of June 1884 and began out operating nearly right away on the Device Works on Manhattan's Lower East Side, an overcrowded keep with a labor stress of a few hundred machinists, humans, manage frame of human beings, and twenty "region engineers" charged with constructing the town's big electric powered powered energy. Tesla changed into running on debugging setups and enhancing generators, without a doubt as it changed into in Paris. Tesla may also moreover want to have most effective met employer creator Thomas Edison some of instances, regular with historiographer W. Bernard Carlson. After being up all night time fixing the harmed eager beavers on the sea liner

SS Oregon, Tesla ran into Batchelor and Edison, who made a statement approximately their "Parisian" being out all night time time time, consistent with Tesla's memoirs. "This is a damned unique man," Edison said to Batchelor after Tesla notified them he had been up all night time time recuperation the Oregon. One of Tesla's aspirations changed into to supply a street lights device primarily based upon arc lighting.

Arc lighting fixtures became the most ordinary kind of avenue lighting fixtures, however it wanted immoderate voltages and end up incompatible with Edison's low-voltage incandescent system, essential to Edison losing agreements in some towns. Tesla's strategies had been in no manner ever taken into production, each due to medical developments in incandescent avenue lighting or a settlement Edison made with an arc lighting fixtures industrial agency for setup.

Tesla had only been at the Device Works for six months at the same time as he decided on to head away.

It's uncertain what delivered approximately him to go away. It could have been over a advantage he did no longer get, either for revamping turbines or for the deserted arc lights plan. Tesla had past squabbles with Edison over unsettled perks he belief he sincerely deserved.

The supervisor of the Edison Device Works provided Tesla a $50,000 benefit to growth "twenty-four remarkable ranges of essential gadgets," regular with Tesla's memoirs, "however it ended up being a prank." According to later variations of the tale, Thomas Edison furnished and after that broke the deal, announcing, "Tesla, you do no longer recognise our American humor."

Since Device Works boss Batchelor emerge as conservative with pay and the commercial enterprise corporation did now not have that quantity of coins on hand (equal to $1.Four million in 2021, the quantity of the reward in each situation has been said as bizarre.

Tesla's journal handiest has one entry approximately what taken region on the prevent of his artwork, a letter written all through 2 pages from December seventh, 1884, to January fourth, 1885, reading "Farewell to the Edison Device Works."

Tesla end up running on patenting an arc lights device rapid after leaving the Edison enterprise corporation, possibly the first-rate equal one he had created at Edison.

In March 1885, he met Edison's patent attorney, Lemuel W. Serrell, to get help with filing the patents.

Serrell led Tesla to two financiers, Robert Lane and Benjamin Vail, who consented to cash the Tesla Electric Light & Production, an arc lights production and energy organization in Tesla's call.

Tesla spent the the relaxation of the 365 days pursuing patents, alongside facet however now not limited to the primary patents given to Tesla in the United States, and manufacturing and putting in region the device in Rahway, New Jersey. The technical press paid interest to Tesla's new tool, applauding its outstanding capabilities.

Tesla's put together for emblem spanking new sorts of rotating modern-day vehicles and electrical transference devices drew little interest from financiers. After the energy become advanced within the one year 1886, they felt that the manufacturing part of the enterprise changed into too competitive and decided on to cognizance best on providing electric powered power. They superior a ultra-current strength agency, eliminating Tesla's and leaving him poverty-bothered. Tesla even out of place custody of his very own patents, which he had passed to the business enterprise in change for stock. He had to artwork for $2 each day at severa electric powered repair art work jobs and as a ditch digger.

Chapter 9: The Induction Motor And Ac

Tesla met Alfred S. Brown, a Western Union superintendent, and Charles Fletcher Peck, a New york metropolis legal professional, in past due 1886.

The 2 guys had preceding experience forming groups and using improvements and patents for monetary benefit. They promised to help Tesla economically and manage his patents primarily based totally actually upon his new electric powered tool mind, and that includes a thermo-magnetic motor concept. In April 1887, they set up the Tesla Electric Company, with an affiliation that 13 percent of profits from patents created may additionally need to go to Tesla, 13 percent to Peck and Brown, and 13 percent to development. At 89 Liberty Street in Manhattan, they superior a lab for Tesla, in which he labored on updating and growing new varieties of electric motors, generators, and different devices.

Tesla advanced a rotating present (AC) induction motor within the yr 1887, a power tool format that have become brief acquiring attraction in Europe and the united states due to its blessings in long-distance, excessive-voltage transference. To turn the motor, polyphase contemporary-day modified into hired, which usual a revolving electromagnetic area (a idea that Tesla said to have conceived inside the 12 months 1882).

This ground-breaking electric motor, created in the month of May 1888, changed into an clean self-starting format that did no longer want a commutator, getting rid of stimulating and the heavy upkeep of frequently fixing and converting mechanical brushes.

In addition to getting the motor patented, Peck and Brown made strategies to promote it it, starting with independent screening to assure that it have become a realistic enhancement, then sending out information release to technical regulars for quantities to run simultaneously with the patent's issuance.

Tesla's AC motor become established in advance than the American Institute of Electrical Engineers on May 16th, 1888, with the aid of physicist William Arnold Anthony (who tested the motor) and Electrical World book editor Thomas Commerford Martin.

Tesla had a accessible AC motor and accompanying energy device, in keeping with Westinghouse Electric & Production Company engineers, which have become a few factor Westinghouse wanted for the rotating current tool he have become already advertising and advertising and marketing. Westinghouse considered claim a patent on a comparable commutator-less, turning magnetic task-based totally induction motor evolved within the 12 months 1885 by way of Italian physicist Galileo Ferraris and released in a paper inside the month of March 1888 thru Tesla, but determined that Tesla's patent should probable manage the market.

In July 1888, Brown and Peck consented to certify Tesla's polyphase induction motor and transformer designs to George Westinghouse for $60,000 in cash and inventory, plus a royalty of $2.50 in line with AC horse power produced through each motor. Westinghouse moreover employed Tesla as an expert on the Westinghouse Electric & Production Company's Pittsburgh lab for a yr for a great settlement of $2,000 ($ 57,six hundred in current day dollars.

Tesla operated in Pittsburgh at the time of that yr, helping inside the improvement of a rotating current system to strength the town's trams. Because of arguments with unique Westinghouse engineers over the manner to first-rate execute AC power, he concept it to be an attempting duration. They decided on Tesla's endorsed 60-cycle AC gadget (to healthy the useful frequency of Tesla's motor), however they unexpectedly decided that it would not paintings for trams thinking about Tesla's induction motor ought to simplest run at a everyday pace. Rather, they determined on a DC traction motor.

Market Discontent

Tesla's dialogue of his induction motor, and Westinghouse's following licensing of the patent, each happened inside the twelve months 1888, at a period while electrical services were in intense competition.

Westinghouse, Edison, and Thomson-Houston had been all seeking to develop in a capital-full-size marketplace at the same time as destructive some different economically. Edison Electric even ran a "conflict of currents" advertising and marketing and marketing try, pronouncing that their direct present device come to be higher and further stable than Westinghouse's rotating gift tool.

Because Westinghouse might be contending in this marketplace, it'd not have the monetary or engineering assets to offer Tesla's motor and related polyphase tool proper now.

Westinghouse Electric changed into in risk 2 years after signing the Tesla settlement. The near-fall apart of London's Barings Bank inspired the 1890 monetary panic, prompting financiers to touch loans to Westinghouse Electric. The agency grow to be compelled to re-finance its loans because of an sudden liquidity lack.

The new consumers requested that Westinghouse lessen down on what seemed outrageous fee on acquisitions, studies, and patents, and that consists of the Tesla settlement's consistent with-motor royalty.

The Tesla induction motor had failed and end up stuck in development at the time. Westinghouse changed into paying an ensured royalty of $15,000 a three hundred and sixty five days , no matter the reality that practical examples of the motor were confined, and the polyphase power structures had to run it were an awful lot rarer.

In early 1891, George Westinghouse informed Tesla approximately his financial problems in blunt terms, specifying that if he did no longer meet his lending institutions' desires, he must lose control of Westinghouse Electric and Tesla would possibly want to "deal with the creditors" to acquire destiny royalties. The blessings of getting Westinghouse keep to sell the motor had been in all likelihood obvious to Tesla, and he consented to let the agency out of the agreement's royalty price state of affairs.

As a part of a patent-sharing control General Electric, Westinghouse bought Tesla's patent for a flat quantity charge of $216,000 6 years later (a business enterprise made from the 1892 merger of Edison and Thomson-Houston).

Chapter 10: The Labs In New York

Tesla ended up being for my part wealthy due to the cash he crafted from accrediting his air conditioning patents, giving him the time and belongings to chase after his very private hobbies. Tesla vacated the Liberty Street shop Peck and Brown had leased inside the three hundred and sixty five days 1889 and labored out of a chain of workshop/laboratory web web sites in Manhattan for the following twelve years. A lab at a hundred 75 Grand Street (1889-- 1892), the 4th and fifth tiers of 33-- 35 South Fifth Opportunity (1892-- 1895), and the sixth and 7th floorings of forty six & 48 East Houston Street (1895-- 1902) have been among them. In the ones workshops, Tesla and his labored with humans did a number of their most crucial art work.

Tesla's Coil

Tesla went to the 1889 Exposition Universelle in Paris inside the summer time of 1889, wherein he decided Heinrich

Hertz's 1886-- 1888 experiments displaying the presence of electro-magnetic radiation, and that includes radio waves.

Tesla concept this new discovery have become "rejuvenating" and solved to have a look at it farther. Tesla attempted powering a Ruhmkorff coil with a excessive-speed generator he changed into growing as part of an extra arc lighting device but decided that the excessive-frequency gift overheated the iron middle and melted the insulation a number of the precept and secondary windings in the coil on the same time as he repeated and broadened on these experiments. To resolve this problem, Tesla developed his "oscillating transformer," which had an air place the various primary and secondary windings in place of insulating cloth, and additionally an iron center that is probably pushed inside and out of the coil. The Tesla coil, because it have become in a while seemed, come to be used to create excessive-voltage, low-modern-day, immoderate-frequency alternating-contemporary-day electricity. This resonant transformer circuit ought to turn out to be utilized in his cordless power research.

Lighting that isn't always confused

After 1890, Tesla tampered with the use of excessive a/c voltages created by means of way of using his Tesla coil to keep electrical electricity thru inductive and capacitive coupling.

He tried to assemble a cordless lighting fixtures gadget primarily based absolutely upon near-project inductive and capacitive coupling by manner of the usage of lighting up Geissler tubes or maybe incandescent moderate bulbs from at some stage in a section in a sequence of public shows. He spent the majority of a whole decade running on wonderful varieties of this new shape of lights with the useful resource of many financiers, however now not one of the obligations succeeded in turning his discoveries into an commercial item.

Tesla suggested observers inside the 3 hundred and sixty 5 days 1893 in St. Louis, Missouri, Philadelphia, Pennsylvania, and the National Electric Light Association that he end up powerful that a gadget like his have to ultimately perform "intelligible signs or perhaps even energy to any distance with out the application of wires" via performing it thru the Earth.

Tesla have become a vice-president of the American Institute of Electrical Engineers, the progenitor of the modern-day IEEE, from 1892 to 1894. (together with the Institute of Radio Engineers).

The Columbian Exposition and the Polyphase System

Westinghouse engineer Charles F. Scott and after that Benjamin G. Lamme had made paintings on an powerful model of Tesla's induction motor by using the use of the begin of 1893. By growing a rotary converter, Lamme had the capability to make the polyphase device it desired suitable with earlier unmarried-phase air con and DC systems. With a manner to provide electricity to all viable customers, Westinghouse Electric began marketing their polyphase air conditioner system as the "Tesla Polyphase System." Tesla's patents, they clearly believed, gave them patent top precedence over brilliant polyphase aircon structures.

Westinghouse Electric welcomed Tesla to reveal within the 1893 World's Columbian Exposition in Chicago, in which the business enterprise had a big space devoted to electric powered exhibitions within the "Electrical energy Building." The quote to slight the Exposition with rotating current modified into acquired via Westinghouse Electric, and it changed into a watershed second in the information of air conditioner energy, due to the reality the enterprise

enterprise confirmed to the American public the security, dependability, and overall performance of a polyphase rotating present device that could additionally energy the alternative air conditioning and DC suggests on the trustworthy.

Tesla's induction motor grow to be established in many shapes and brands in a completely unique exhibition hall. A sort of shows were used to expose the spinning electromagnetic issue that moved them, which incorporates an Egg of Columbus that employed a -segment coil decided in an induction motor to spin a copper egg, triggering it to stand on stop.

Throughout the honest's six-month run, Tesla stopped via for each week to go to the International Electrical Congress and do a series of shows at the Westinghouse form.

Tesla showed his cordless lighting fixtures machine in an deliberately darkish chamber, the usage of a presentation he had previously accomplished throughout America and Europe, which comprised using immoderate-voltage, excessive-frequency rotating gift to mild cordless gas-discharge lighting fixtures.

Oscillating Generator Powered by manner of the usage of Steam

Tesla delivered his steam-powered reciprocating electric powered energy generator, which he patented that three hundred and sixty five days, inside the course of his dialogue on the International Electrical Congress in the Columbian Exposition Farming Hall, which he concept end up a far higher way to provide rotating present.

Steam located into the oscillator and out via a sequence of ports, requiring a piston combined to an armature up and down. A rotating electromagnetic challenge come to be created via the magnetic armature vibrating up and down at a immoderate price. The surrounding wire coils skilled rotating electric present due to this. It removed the tough factors of a steam engine/generator, however it have become in no way ever taken into consideration a likely engineering selection for generating electric electricity.

Niagara Consulting

In the year 1893, Edward Dean Adams, the president of the Niagara Falls Cataract Building and introduction Company, approached Tesla for steering at the gold stylish tool for shifting electricity created at the falls. There has been a chain of mind and open competition for some years at the way to efficaciously harness the energy supplied via way of the falls. Two-segment and three-phase air con, excessive-voltage DC, and compressed air had been a number of the options supplied with the resource of many USA and European agencies. Adams inquired on the prevailing kingdom of all contending structures from Tesla. Tesla cautioned Adams that a -phased system would be the maximum respectable, and that a Westinghouse gadget for lights incandescent lighting fixtures with -segment rotating gift existed. Based upon Tesla's tip and Westinghouse's presentation at the Columbian Exposition that they might boom an extensive a/c gadget, the agency gave an settlement to Westinghouse Electric to construct a -segment air conditioning growing machine on the Niagara Falls. General Electric have turn out to be moreover granted an settlement to construct the air con move device at the precise same time.

Nikola Tesla's Business

In the 365 days 1895, Edward Dean Adams consented to assist form the Nikola Tesla Company, that could cash, expand, and marketplace quite a few in advance Tesla patents and improvements, and new ones, after exploring Tesla's laboratory. Alfred Brown signed up with the business enterprise, bringing with him patents created through Peck and Brown. William Birch Rankine and Charles F. Coaney had been contributed to the board. Company delivered in few financiers due to the fact the mid-1990s had been a tough monetary period, and the patents for cordless lighting fixtures and oscillators that it modified into installation to market never ever emerged. Tesla's patents were controlled with the aid of the use of the enterprise for many years.

The Fire within the Laboratory

The South Fifth Opportunity structure that housed Tesla's laboratory ignited inside the morning hours of March thirteenth, 1895. It began out in the form's basement and ended up being so immoderate that Tesla's fourth-story laboratory charred and crashed into the second floor. Tesla's continuous art work were hindered by using using the fireplace, which moreover damaged a fixed of early notes and research observe materials, fashions, and presentation portions, masses of which have been tested on the 1893 Worlds Colombian Exposition.

The New York Times said on Tesla's feedback "I am in manner too much ache to talk. I am uncertain what to say." Tesla transferred to forty six & 48 East Houston Street after the fireside and restore his laboratory on the 6th and 7th floorings.

Experiments the use of X-Rays

After seeing broken film in his lab on the time of preceding checks, Tesla began exploring what he defined as glowing radiation of "unnoticeable" paperwork inside the one year 1894. (later decided as "Roentgen rays" or "X-Rays"). Crookes tubes, a cold cathode electrical discharge tube, were his first examinations. When Tesla tried to envision Mark Twain lit up with the beneficial useful resource of a Geissler tube, an early form of gas discharge tube, he should have accidently gotten an X-ray picture, preceding through manner of some weeks Wilhelm Röntgen's December 1895 announcement of the invention of X-rays.

The steel locking screw on the electronic camera lens modified into the fine factor caught inside the shot.

After becoming aware of Röntgen's discovery of X-rays and X-ray imaging (radiography) within the month of March 1896, Tesla set out to perform his own experiments in X-ray imaging, developing a excessive-electricity unmarried-terminal vacuum tube without any goal electrode and powered with the useful resource of the Tesla Coil's output (the cutting-edge term for the phenomenon produced thru this device is bremsstrahlung or braking radiation). Tesla created many speculative setups to supply X-rays as part of his studies research. Tesla stated that his circuits would probable allow him to "produce Roentgen rays of some distance more pressure than is feasible with regular device."

Tesla suggested approximately the threats of dealing with his circuit and unmarried-node X-ray generators. He ascribed the pores and pores and skin harm to many motives in his massive notes on the early assessment of this problem. He felt that pores and skin harm modified into triggered with the resource of ozone usual in touch with the pores and skin, and to a lower level, nitrous acid, in vicinity of thru Roentgen rays.

Tesla come to be misinterpreted in wondering that X-rays were longitudinal waves just like the ones produced with the aid of plasma waves.

Those plasma waves can take area in electromagnetic fields with little pressure.

Push-button Control for Radio

Throughout an electrical exposition at Madison Square Garden inside the three hundred and sixty five days 1898, Tesla confirmed to the general public a ship that hired a coherer-primarily based absolutely radio manage, which he known as "telautomaton."

Tesla attempted to market his innovation as a form of radio-controlled torpedo to the U.S. Military, however they had been withdrawn.

Remote radio manage stayed a novelty until WW1 and later, when innovation was launched in military programs via tremendous international locations. In his cope with to the Business Club in Chicago on May the 13th 1899, Tesla used the hazard to farther display "Teleautomatics" at the identical time as on his manner to Colorado Springs.

Chapter 11: The Principle Of Wireless Energy

Tesla committed the majority of his money and time from the Nineties to 1906 working on a succession of experiments focused at growing the switch of electrical energy with out the utility of wires. It grow to be an development of his coil-based electricity transference precept, which he had previously confirmed in cordless lighting. He imagined this as an technique to switch no longer top notch massive quantities of electricity everywhere within the worldwide, however moreover, as he had noted in earlier lectures, round the arena interactions.

There grow to be no possible way to wirelessly supply out conversation symptoms over massive distances, now not to say huge portions of electrical power, while Tesla modified into forming his ideas. Tesla had investigated radio waves early on and concluded that a number of Hertz's former studies on them emerge as wrong.

Also, at the time, this new type of radiation grow to be normally idea to be a quick-range phenomenon that seemed to disappear in a lot much less than a mile. Even if radio wave thoughts were proper, Tesla confused that they have been useless for his dreams due to the fact this kind of "unnoticeable mild" ought to decrease with time, like some different radiation, and could adventure in straight away strains out into region, being "hopelessly out of place."

Tesla started out running on assessments to check the speculation that he could have the ability to perform electric powered electricity over massive distances through the Earth or environment in the mid-Eighteen Nineties, and that includes installing vicinity an significant resonance transformer amplifying transmitter in his East Houston Street laboratory.

He proposed a gadget which consist of balloons postponing, transferring, and getting electrodes in the air above 30,000 toes (9,100 m) in pinnacle, wherein he virtually believed the lower pressure ought to permit him to move extremely good voltages (endless volts) over u . S . Miles.

Colorado Springs

Tesla advanced a speculative station at a immoderate elevation in Colorado Springs within the 12 months 1899 to extra check the conductive attributes of low-stress air.

He want to safely run substantially bigger coils there than he need to in his New York laboratory, and a companion had scheduled the El Paso Power Company to offer rotating gift freed from charge. To resource his experiments, he encouraged John Jacob Astor IV to become being a bulk investor inside the Nikola Tesla Company for $one hundred,000 ($ 3,a hundred and ten,800 in cutting-edge bucks. Astor honestly believed he have become specifically putting coins proper into a modern-day cordless lighting fixtures device. Tesla, however, used the cash to cash his experiments in Colorado Springs.

He advised press newshounds on his visit that he presupposed to perform cordless telegraphy experiments, sending out indicators from Pikes Peak to Paris.

There, he tampered with an big coil strolling inside the megavolt variety, producing synthetic lightning (and thunder) with endless volts and discharges as a great deal as a hundred thirty five ft (forty one m) in length, and incorrectly forced out the generator in El Paso, main to a strength interruption.

He incorrectly concluded that he need to use the complete worldwide of the Earth to perform electric powered power based totally absolutely upon his observations of the digital sound of lightning actions.

Tesla were given normal signals from his receiver on the time of his time in his lab, which he presumed have been messages from a few different international. In a letter to a press reporter in the month of December 1899 and a letter to the Red Cross Society within the month of December 1900, he referred to them.

Press reporters jumped to the idea that Tesla changed into getting indicators from Mars because it turned into a extraordinary story. In a Collier's Weekly brief article entitled "Talking with Planets" on February ninth, 1901, he elaborated on the messages he heard, announcing that it wasn't at once obvious to him that he changed into getting "smartly directed interactions" and that the signs may additionally have originated from Mars, Venus, or distinctive worlds. Tesla need to have obstructed Guglielmo Marconi's European tests in the month of July 1899-- Marconi could have sent out the letter S (dot/dot/dot) in a marine presentation, the best same three impulses Tesla hinted about paying attention to in Colorado -- or alerts from a few different cordless experimenter.

Tesla had negotiated with The Century Publication's editor to write down down a quick article about his consequences. The book dispatched a expert photographer to Colorado to record the paintings in development. The essay, entitled "The Issue of Increasing Human Energy," changed into released in the e-book's June 1900 trouble. He defined why the cordless tool he pictured transcended, even though the piece emerge as extra of a long philosophical writing than an available medical account of his artwork, accompanied with what would possibly grow to be being well-known snap shots of Tesla and his experiments in Colorado Springs.

Wardenclyffe

Tesla visited New York seeking out financiers for what he in truth believed is probably an powerful cordless transference tool, amusing them on the Waldorf-Palm Astoria's Garden (in which he have become very last at the time), The Players Club, and Delmonico's ingesting institutions.

He had been given $one hundred fifty,000 ($4,666,two hundred in current greenbacks from J. P. Morgan inside the month of March 1901 in exchange for a fifty one percent stake of any cordless patents created, and started making plans the Wardenclyffe Tower website in Shoreham, NY, one hundred miles (161 km) east of the metropolis on Long Island's North Coast.

Tesla had multiplied his intents to boom a greater effective transmitter via manner of July 1901, to conquer Marconi's radio-based totally device, which Tesla mistook for his personal.

Morgan reduced to provide any delivered financing whilst he approached him to look for cash to enlarge the bigger tool. Marconi efficaciously communicated the letter S from England to Newfoundland inside the month of December 1901, exceeding Tesla within the competition to be the first to acquire this type of transference. Tesla tried to influence Morgan to fund a honest large proposition to relay interactions and electric energy by using controlling "vibrations over the world" a month after Marconi's success. Tesla wrote more than 50 letters to Morgan over the next five years, attractive for and demanding brought coins to emerge as Wardenclyffe. In the yr 1902, Tesla endured operating at the undertaking for another nine months. The tower became raised to its maximum powerful 187-foot top (fifty seven m). Tesla transferred his laboratory from Houston Street to Wardenclyffe in the month of June 1902.

Wall Street financiers had been putting cash into Marconi's system, and journalism started to assault Tesla's innovation, pronouncing it turn out to be a rip-off.

In the one year 1905, the challenge got here to a grinding halt, and inside the year 1906, monetary issues and one-of-a-kind situations may additionally need to have introduced to Tesla's intellectual disintegrate, regular with Tesla chronicler Marc J. Seifer. Tesla used the Wardenclyffe belongings as protection to cover his Waldorf-Astoria monetary duties, which amounted to $20,000 ($516,700 in modern-day greenbacks. In the year 1915, he out of place the property to foreclosure, and the modern owner took apart the Tower within the yr 1917 to make the net website on-line a extra precious assets belongings.

Chapter 12: Other Events And Developments

To limit the flow of facts among the 2 international locations, the British detached the transatlantic telegraph connection linking the U.S. And Germany even as The First World War broke out. They furthermore tried to cut off German cordless connection to and from the united states through method of getting the united statesA. Marconi Company prosecute Telefunken for patent infraction. For their defense, Telefunken worked with physicists Jonathan Zenneck and Karl Ferdinand Braun, and Tesla, who emerge as paid as a witness for 2 years at $1,000 month-to-month. When the us went into the warfare in opposition to Germany within the twelve months 1917, the claim suffered and ultimately ended up being moot.

Tesla attempted to take legal action in opposition to Marconi for violation of his cordless tuning patents inside the one year

1915. Marconi's first radio patent modified into launched in the US inside the 365 days 1897, but his 1900 patent software program software for enhancements to radio transference emerge as continuously have turn out to be down in advance than being criminal within the 12 months 1904 at the premises that it infringed on extraordinary patents, inclusive of but not limited to two Tesla cordless power tuning patents launched within the one year 1897. Tesla's 1915 case failed, however a Supreme Court of the usa judgment in the 365 days 1943 delivered once more the sooner patents of Oliver Lodge, John Stone, and Tesla in an related case wherein the Marconi Company attempted to take prison movement in competition to the united statesA. Federal authorities over WWI patent violations. The court docket mentioned that its judgment had no impact on Marconi's claim to being the primary to carry out radio transference; as an alternative, sincerely due to the truth Marconi's declare to fine trademarked upgrades doubted, the corporation couldn't claim violation on those patents.

The Nobel Reward in Physics become rewarded to Thomas Edison and Nikola Tesla on November sixth, 1915, in keeping with a Reuters information business organisation corporation report from London; regardless of the truth that, on November fifteenth, a Reuters tale from Stockholm referred to that the praise that year come to be being rewarded to William Henry Bragg and Lawrence Bragg "for their services inside the evaluation of crystal shape through manner of manner of X-rays."

At the time, there had been unproven claims that each Tesla or Edison had rejected the award. "Any fable that an character has now not been given a Nobel Reward sincerely because of the truth he has made appeared his goal to refuse the benefit is ridiculous," in step with the Nobel Structure; a recipient can best decrease a

Nobel Reward after he has been confirmed a winner.

Tesla biographers have said that Edison and Tesla were the actual receivers, and that neither become given the award due to their bitterness closer to every precise; that each seemed to lower the possibility's accomplishments and right to win the award; that both did not want to accept the award if the possibility have been given it to start with; that each became down any possibility of sharing it; and that even a wealthy Edison refused it to hold Tesla from getting the $20,00 reward.

Tesla and Edison did not win the reward within the years following these critiques (regardless of the truth that Edison were given one in each of 38 feasible costs within the year 1915 and Tesla have been given one in every of 38 viable costs inside the twelve months 1937).

Tesla attempted to sell some gadgets based totally totally absolutely upon ozone production. His 1900 Tesla Ozone Company, for instance, presented an 1896 trademarked tool primarily based upon his Tesla Coil, which bubbled ozone through numerous oils to supply a medical gel. Several years later, he attempted to increase a version of this as a sanatorium room sanitizer.

The software of electrical strength to the thoughts, constant with Tesla, prolonged intelligence. In the yr 1912, he developed a ""a way to make silly youngsters brilliant with the useful useful resource of with the resource of coincidence saturating them with electric powered powered strength," electrical wiring a schoolroom's walls, and "saturating [the schoolroom] with tiny electrical waves oscillating at high frequency." Mr. Tesla assures that the whole region might be modified proper into a fitness-giving and revitalizing electromagnetic difficulty or 'tub.'" The precept come to be licensed, as a minimum provisionally, through manner of William H. Maxwell, the superintendent of New York City schools on the time.

Tesla searched for distant places financiers earlier than World War I. Tesla misplaced the coins he grow to be obtaining from his patents in European international places after the battle broke out.

Tesla proposed in the August 1917 hassle of the e-book Electrical Experimenter that electrical energy might be used to decide submarines by the use of using displaying a "electric beam" of "immoderate frequency" and seeing the be part of up a fluorescent display display screen (a device that has been said to have a shallow similarity to trendy-day radar).

Tesla's idea that immoderate-frequency radio waves may also moreover additionally penetrate water showed wrong. Tesla's wellknown speculation that a certainly robust immoderate-frequency signal might be wished held true, constant with Émile Girardeau, who helped expand France's first radar device inside the Nineteen Thirties. "(Tesla) end up prophesying or dreaming, just due to the reality he had no manner of sporting them out," Girardeau mentioned, "however one need to add that if he modified into dreaming, he was dreaming correctly."

Tesla's ultimate patent, USA Patent 1,655,114, become provided in the 12 months 1928 for a biplane inexperienced in vertical liftoff and touchdown (VTOL) that "slowly slanted thru control of the elevator systems" in flight until it flew like a habitual plane.

Tesla approximated that the plane should retail for underneath $1,000.

In spite of being categorised as impractical, the airplane bears early similarities to the

U.S. Armed strain's V-22 Osprey. Tesla's staying workplace at 350 Madison Opportunity, which he had relocated to two years previous, became closed currently.

Chapter 13: Tesla's Death

Tesla departed the Hotel New Yorker after midnight one night time inside the fall of 1937, at the age of 80 one, to make his ordinary trek to the cathedral and library to feed the pigeons. Tesla emerge as tossed to the ground at the same time as crossing a road a few blocks from the lodge after failing to maintain from a shifting taxicab. In the accident, his again have turn out to be badly twisted, and three of his ribs have been fractured. Tesla's conditions were in no manner ever in reality comprehended due to the fact that he refused to peer a scientific doctor, a exercise he had practiced for essentially his entire lifestyles, and he by no means ever clearly recuperated.

Tesla passed away by myself in Space 3327 of the Hotel New Yorker on January seventh, 1943, at the age of 86. After going into Tesla's room no matter the "do not disrupt" indication Tesla had located on his door 2 days preceding, house maid Alice

Monaghan positioned his loss of existence. H.W. Wembley, an assistant health inspector, analyzed the frame and decided that coronary apoplexy was the reason of death.

The Alien Property Custodian come to be ordered through the Federal Bureau of Investigation to take Tesla's subjects 2 days later.

John G. Trump, an M.I.T. Professor and popular electric engineer who works as a technical assistant to the National Defense Research Study Committee, became summoned to check the Tesla merchandise in custody.

www.ingramcontent.com/pod-product-compliance
Lightning Source LLC
Chambersburg PA
CBHW071444080526
44587CB00014B/1990
* 9 7 8 1 7 7 7 0 6 6 3 5 2 *